PRAISE FOR

The Zero Waste Solution

"Paul Connett is one of the few people I know who can make waste interesting. In this book, he glides you through the diverse world of garbage and guides you to the many, ready, efficient, and safe solutions that don't involve you breathing in the toxics of the incinerator industry."

—RALPH NADER, consumer advocate;
author, *The Seventeen Solutions*

"In this era of mounting environmental problems, we need more spotlights on solutions—not theoretical ones, but real, on-the-ground ones we can start implementing in our communities today. *The Zero Waste Solution* is just what the doctor ordered! Packed with inspiring case studies and helpful guidance, Connett's book lays out a roadmap of how to turn our wasteful consumer lifestyle into one of sustainability and health. Read this book and join the movement for real solutions!"

—ANNIE LEONARD, author, *The Story of Stuff*

"*The Zero Waste Solution* is for all those concerned about humanity's health and environment. Essential reading for anyone fighting landfills, incineration, overpackaging, and the other by-products of our unthinking and irresponsible throwaway society."

—JEREMY IRONS, actor;
executive producer, *Trashed*

"Paul Connett's heroic drive to trim our culture's crazy waste has helped produce the powerful movement chronicled here—an inspiration to us all!"

—BILL MCKIBBEN, author,
Oil and Honey: The Education of an Unlikely Activist

"Connett's book should inspire everyone personally, citizens' groups, and all levels of government to reduce waste—an insidious problem engulfing the world."

—THEO COLBORN, president, TEDX
(The Endocrine Disruption Exchange);
professor emeritus, University of Florida;
honorary professor of science, University of Colorado;
and lead author, *Our Stolen Future*

"Connett's book correctly illuminates the essential nature of citizen-led movements for true zero waste solutions to transform society. Using Connett's methods, the It's Not Garbage Coalition in Nova Scotia stopped a string of imported garbage incinerators, a PCB incinerator, a medical-waste incinerator, a massive MSW incinerator, and a sewage-sludge incinerator. By 1994 we had a provincewide ban on incineration and a citizen-designed, world-leading system aimed toward zero waste. With the wisdom in this book, you can achieve results like this, too!"

—DAVID WIMBERLY, cochair, It's Not Garbage Coalition

"*This* would have been the book to have when we first began fighting the incinerator proposal in our region of Ontario. It's full of relevant, factual information and suggests specific questions to pose to decision makers and regulators. It also demonstrates that a sustainable zero waste strategy is achievable, and far more preferable than society's current focus on waste disposal."

—KERRY MEYDAM AND WENDY BRACKEN,
Durham Environment Watch, and
LOUIS BERTRAND AND LINDA GASSER,
Zero Waste 4 Zero Burning, Ontario

THE
ZERO
WASTE
SOLUTION

THE ZERO WASTE SOLUTION

Untrashing the Planet One Community at a Time

PAUL CONNETT

CHELSEA GREEN PUBLISHING

WHITE RIVER JUNCTION, VERMONT

Editor: Joni Praded
Project Manager: Patricia Stone
Copy Editor: Eric Raetz
Proofreader: Alice Colwell
Indexer: Shana Milkie
Designer: Melissa Jacobson

Printed in the United States of America.
First printing October, 2013.
10 9 8 7 6 5 4 3 2 1   13 14 15 16 17

Chelsea Green is committed to preserving ancient forests and natural resources. We elected to print this title on 30% post-consumer recycled paper, processed chlorine-free. As a result, we have saved:

22 Trees (40' tall and 6-8" diameter)
10 Million BTUs of Total Energy
1,943 Pounds of Greenhouse Gases
10,540 Gallons of Wastewater
705 Pounds of Solid Waste

Chelsea Green made this paper choice because our printer, Thomson-Shore, Inc., is a member of Green Press Initiative, a nonprofit program dedicated to supporting authors, publishers, and suppliers in their efforts to reduce their use of fiber obtained from endangered forests.

For more information, visit www.greenpressinitiative.org

Environmental impact estimates were made using the Environmental Defense Paper Calculator. For more information visit: www.edf.org/papercalculator

Our Commitment to Green Publishing

Chelsea Green sees publishing as a tool for cultural change and ecological stewardship. We strive to align our book manufacturing practices with our editorial mission and to reduce the impact of our business enterprise in the environment. We print our books and catalogs on chlorine-free recycled paper, using vegetable-based inks whenever possible. This book may cost slightly more because it was printed on paper that contains recycled fiber, and we hope you'll agree that it's worth it. Chelsea Green is a member of the Green Press Initiative (www.greenpressinitiative.org), a nonprofit coalition of publishers, manufacturers, and authors working to protect the world's endangered forests and conserve natural resources. *The Zero Waste Solution* was printed on FSC®-certified paper supplied by Thomson-Shore that contains at least 30% postconsumer recycled fiber.

Library of Congress Cataloging-in-Publication Data
Connett, P. H. (Paul H.)
  The zero waste solution : untrashing the planet one community at a time / Paul Connett.
    pages cm
  Includes bibliographical references and index.
  ISBN 978-1-60358-489-0 (pbk.)—ISBN 978-1-60358-490-6 (ebook)
1. Source reduction (Waste management) 2. Recycling (Waste, etc.) 3. Refuse and refuse disposal. 4. Pollution prevention. I. Title.

TD793.95.C66 2013
628.4—dc23

                                    2013025260

Chelsea Green Publishing
85 North Main Street, Suite 120
White River Junction, VT 05001
(802) 295-6300
www.chelseagreen.com

MIX
Paper from
responsible sources
FSC® C013483

CONTENTS

CONTRIBUTORS

Paul Connett, author

WITH GUEST CHAPTERS OR ESSAYS BY:
Cecilia Allen
Katy Anderson
Richard Anthony
Buddy Boyd
Mary Lou Van Deventer
Sonia Dias
Rossano Ercolini
Froilan Grate
Daniel Knapp
Gary Liss
Patrizia Lo Sciuto
Eric Lombardi
Jeffrey Morris
Neil Seldman
Bill Sheehan
Joan Marc Simon
Tommaso Sodano
Helen Spiegelman

AND IMPORTANT CONTRIBUTIONS FROM:
Magdalena Donoso, Virali Gokaldas, Gigie Cruz, Ananda Tan, Neil Tangri,
and others from the Global Alliance for Incinerator Alternatives (GAIA)
as well as Jutta Gutberlet, Laila Iskander, Sonia Mendoza,
and Paul Martensson

FOREWORD

Until three years ago I, like most people, was blissfully unaware of the damage to our health and environment that the trash produced by our consumer society was causing. It was only when director Candida Brady suggested to me that this problem should be the subject of a documentary film, and began to share with me her research, that I began to understand the magnitude of what to most people remains invisible. While filming what became the documentary *Trashed*, we traveled much of the world to record the effects of different societies' attempts to get rid of its garbage, whether by burning it, burying it, dumping it in the water, or just ignoring it.

The evidence was startling, and I soon realized that the massive growth since World War II of global production, consumerism, and its attendant prosperity, for many, had come at a price. As we gloried in our inventiveness and ingenuity at converting the earth's resources into products that made life easier and more enjoyable for so many, it seems we had not addressed the problem of sustaining and truly valuing our world's inevitably finite resources. The health of our air, of our ground water, and of our oceans was something that we assumed nature, in her generous way, would take care of for us. But it became increasingly clear that while nature will take care of herself by adapting, those adaptations might not be to our liking. It is no skin off her nose if sea levels rise, if the weather becomes more violent, if the seasons change, even if the human race becomes sicker, hungrier, thirstier, or even extinct. She will continue on regardless.

I hope we are nearing the end of what I have come to recognize as our consumer adolescence—seventy years during which we have run riot with our newly discovered abilities without having yet learned the responsibilities that these newfound powers incur. Only when we reach maturity do we begin to think of our children and grandchildren, and of the world we are responsible for passing on to them. It is the burgeoning of this more mature attitude that has given me such hope as I have traveled presenting *Trashed* to audiences worldwide. It has become clear to me there is a huge body of people committed to protecting our health and our environment, and to doing what is necessary to deal with the mountains of trash that we waste. But it has also become clear to me that such behavior needs continuing personal effort. Individuals need to make daily decisions to separate their garbage intelligently, and pressure their government representatives to put a collection system in place that will allow their trash to be turned into compost or reused in future production.

Our governments also, both local and national, must play their part by putting systems in place so that our trash can become an economic resource, rather than simply a problem. And they must be held accountable if they try to convince us that burning it will produce cheap electricity, and burying it will provide cheap gas. It takes twenty-six times more energy to incinerate the ubiquitous plastic water bottle than it takes to produce one in the first place. That to me seems like waste, not "waste to energy."

Industry also needs to become more transparent and to play its part, and there are heartening examples of this beginning to happen. Many responsible companies are investing to find ways of producing goods without relying on toxic chemicals. They are also allowing their products to be returned when no longer in use, so that the precious metals and other components can reused in new products.

But it is you and I who can help encourage this attitude, starting with ensuring that what we buy is recyclable and has been produced in a sustainable way. The market is *us*, and if we demand certain things and certain behaviour from the manufacturing industry, then it will be very much to its advantage to comply.

First of all, though, we need to understand fully the problems that our garbage creates. And then we need to learn the solutions for dealing with those problems from the people who have researched them. And this is where Paul Connett's remarkable book becomes invaluable. Paul is perhaps *the* world expert on the problems and solutions of garbage reduction and reuse, and his book and its contained wisdom will, I hope, become compulsory reading for anyone whose home is threatened by plans to build incinerators, or to site landfills, or who cares about leaving this world in a fit state for future generations. I am enormously grateful for his help with *Trashed*, and to his ceaseless efforts to keep the subject of garbage and the wasting of the world's resources higher on the agenda than it otherwise might be.

JEREMY IRONS

PREFACE

Unexpectedly, waste management has occupied twenty-eight years of my life. It has taken me around the world. Along the way I've made many, many friends at the grassroots level who have been my greatest teachers about commonsense solutions to the waste problem. For me, eliminating waste entirely, and dealing with it safely and sustainably until we do, has become personal.

This is not how I thought I'd spend these past three decades. As John Lennon wrote in "Beautiful Boy," "Life is what happens to you while you're busy making other plans"—a sentiment that sums up my life very well. After graduating from Cambridge University and a short time teaching high school chemistry in England, I went to the United States to embark on a PhD program in biochemistry at Cornell University. In 1967, I was one year into this program when I became embroiled in the Vietnam peace movement. That involvement led to further involvements in the McCarthy for President campaign, the Lowenstein for Congress campaign, the American Committee to Keep Biafra Alive, Action Bangladesh, the Free JP Campaign and the UK campaign of People for a Non-Nuclear World. Then in 1979 I found myself back in a PhD program, this time at the chemistry department at Dartmouth College in New Hampshire, and this time married with three children.

When I graduated from Dartmouth in 1983, I started teaching biochemistry at St. Lawrence University in Canton, New York. I thought my campaigning days were over; I was expecting a quiet life. Then in 1985 I was dragged into a campaign to fight a trash incinerator proposed for our county. This eventually led to the formation of Work on Waste USA. My wife, Ellen, ended up editing the newsletter *Waste Not* for thirteen years, and as director I found myself being invited to give presentations to many community groups on the dangers of incineration. So far this mission has taken me to forty-nine states in the United States, seven provinces in Canada, and sixty other countries. It has involved approximately 2,500 talks. I do not charge communities for my services; however, my travel and accommodation are taken care of, and I am amply paid in gratitude—and in many cases lasting friendships with community activists and leaders. I see myself as a consultant working in the community interest on waste management (and some other environmental issues), as opposed to the many highly paid consultants working in the corporate interest.

In 1996, just when I thought I had finally been able to balance the twin demands of teaching chemistry full time at a university and the need to help communities on waste management, Ellen asked me to get involved with fighting water fluoridation. That *little* diversion has lasted another seventeen years and culminated in 2010 in the publication of a book I coauthored with two other scientists, *The Case Against Fluoride: How Hazardous Waste Ended Up in Our Drinking Water and the Bad Science and Powerful Politics That Keep It There* (published by Chelsea Green in 2010).

However, my involvement with the battles over incineration never ended, and that brings us up to the present and to this book. If there is one thing I've learned in years of fighting proposed incinerators, it is that these battles are not just about articulating the dangers and wastefulness of burning trash. Communities only win temporary battles by saying no to something that they can't live with. The long-term victory requires saying yes to something that they can live with. So today most of my time is spent providing information about a cost-effective and comprehensive alternative strategy to both landfills and incinerators. This strategy is called Zero Waste 2020. Not only does this approach provide better economic opportunities at the local level, but it also provides a key stepping-stone toward sustainability at the global level.

Where Can We Find Optimism About the Future?

Sometime in the 1990s Barry Commoner was asked whether he was optimistic about the future; he replied with words to the effect that if citizens were allowed to shape key decisions he was optimistic, but if it was left to governments he was not. As I write these words in 2013 and I think ahead, I am not optimistic about governmental action to move our world toward sustainability. There is much talk but little evidence that we are prepared to take the tough measures that are needed. If this is left in the hands of powerful governments—especially those influenced or controlled by major corporations—I am not optimistic that by the end of the twenty-first century there will be any lifestyle left to which we in industrialized countries have grown accustomed.

Why? Political leaders continually assert that we need more "growth" to get us out of our national and global economic problems. We have become addicted to growth as a universal solution for creating jobs or dealing with national debts or reversing recessions—even though growth as we know it leads to more pollution, more loss of biodiversity, more resource depletion, and more climate change. For nearly half a century, scientists have warned that endless growth on a finite planet is, essentially, a highway to disaster. According to Jorgen Randers, one of the scientists who first issued warnings about the limits to growth, human demand on the biosphere already exceeds the global biocapacity by around 40 percent.

Journalist Chris Hedges has this to say about a world still pinning its hopes on "more growth" to solve its problems:

> This strange twilight moment, in which our experts and systems managers squander resources in attempting to re-create an expanding economic system that is moribund, will inevitably lead to systems collapse. The steady depletion of natural resources, especially fossil fuels, along with the accelerated pace of climate change, will combine with crippling levels of personal and national debt to thrust us into a global depression that will dwarf any in the history of capitalism. And very few of us are prepared.[1]

In fact, more and more voices around the world are emphasizing that endless growth is our problem, not our solution. Political leaders and mainstream economists may be slow to the party, but a growing cadre of everyday people are trying to figure out how to solve problems in new ways.

Among those problems are the mountains of trash that mindless growth spawns. People who are forced to live near landfills or incinerators are already aware of some of the dreadful realities of our global waste problem. To most, it's out of sight and out of mind.

Everyone Is Needed in the Fightback

In promoting the notion of a zero waste society, this book outlines a fightback against the gloomy realities and predictions that characterize our times. The good news is this fightback has the potential to involve everyone from the ragpickers in the Global South to the brightest minds in our universities in the Global North and everyone else in between. Waste landfills, incinerators, and seas riddled with trash provide the visible evidence that our throwaway society simply cannot be sustained on a finite planet. The solution is in everyone's hands once the need to end the throwaway ethic has been forced onto our political leaders' agendas. Meanwhile, many communities around the world— especially in California, Italy, and Spain—are providing working examples of what can be achieved when local leaders are prepared to work with citizens to find waste solutions we can all live with.

So this book invites you to take a look at the "lowly" world of waste. It may seem to be a subject of little importance, something that can be left to others in our busy lives. However, I argue that waste is crucially important because it is something that we all make every day. We make it with our own hands and our own purchasing decisions. If we leave the waste problem to itself, we are part of a nonsustainable way of living on this planet with huge consequences for human health and the global environment. However, with good leadership we can be part of the solution. In a very real sense the future is in our hands. Waste is the place where we *all* connect with sustainability—or rather a lack of it.

Thus, waste is the very best place to start the takeback of our planet and move it a little closer to sustainability. Big problems can arise from ignoring many small bad actions. Elsewhere people have talked about "the tyranny of small decisions." Ultimately it is our own personal responsibility to be part of the first step in the larger movement to move our community or corporation in the right direction. Secure the change here with the more appropriate handling of our discarded objects and materials, and some of our communities and corporations will be ready for some of the other, bigger changes we need to make to move toward a sustainable future.

I do not anticipate that the world will do all the things I advocate in this book right away; however, I am confident that we can set up models that others can copy when the various resource crises make the nonsustainability of our current consumption patterns painfully obvious for all to see. So this is a plan for communities (and some corporations)—from small villages to large cities—who are prepared to be the pioneers for others to follow. We need a blueprint (or several blueprints) for what to do with our "waste" if we wish to move in a sustainable direction.

Fortunately, some communities have already started the approaches outlined in this book and these can be copied right now, and we need many more places to do so. There are many acts—some small and some large—that are happening right now. With the Internet and DVDs these successes can be quickly and readily communicated around the world.

In a different context Alan Paton, author of *Cry, the Beloved Country*, wrote, "There is only one way in which one can endure man's inhumanity to man and that is to try, in one's own life, to exemplify man's humanity to man."[2] In a similar vein, the only way we can endure the world's lack of sustainability is to strive to make our own community an example of a community willing to take steps toward zero waste.

Achieving zero waste is not going to be easy, but like sustainability it is a moral imperative unless we wish to see the complete demise of any semblance of a lifestyle over and beyond the barbaric.

ACKNOWLEDGMENTS

Most of what I know has come from the many people working at the grass-roots level to solve waste problems around the world. They have supported my travels and inspired me with their commonsense solutions. Everywhere in this text that I have written *I* it should read *we*.

A big part of that *we* has been my wife, Ellen. In every campaign I have been involved in since 1968, Ellen has played a major role. She has been the instigator of many of my activities and for twenty-eight years has played a huge role in working on waste issues, especially editing the newsletter *Waste Not* at the height of the battle against incineration in North America from the mid-1980s to the mid-1990s. Again and again in writing this book I have drawn on her huge reservoir of knowledge.

Many others have helped to write this book, and I am especially indebted to Cecilia Allen and many other members of GAIA (the Global Alliance for Incinerator Alternatives). Hopefully, some of GAIA's "case studies" will inspire others to copy the pioneering efforts involved. I would also be very glad to receive accounts of other success stories that we may include in future editions of this text.

I am also indebted to Jack Cook, who assisted me in many important ways with this book. He not only helped to create some of the illustrations in the book and procure others, but also offered his many wise words of encouragement during this effort. In fact without his shoulder to lean on there were times when I might have abandoned this effort completely, as the competition for my time—especially from my anti-fluoridation work—became rather intense.

As anyone who has ever written a book knows, the key person is the editor. I was particularly blessed to have one of the very best: Joni Praded at Chelsea Green. Her skill is the deft ability to rewrite sections to improve the communication but at the same time to leave the message in the author's own voice.

Many of the photographs in this book underline the notion that a picture is worth a thousand words. But to find all the photographs I wanted to use and get permission to use them was a mammoth task. Fortunately I had a great team to help me track them down. This team included Gigie Cruz (Egypt and the Philippines), Enzo Favoino (Italy), Pier Felice Ferri (Italy), Kevin Hurley (Burlington, Vermont), Laila Ikslander (Egypt), Patrizia Lo Sciuto (Italy), Anne Laracas (the Philippines), Eric Lombardi (Eco-Cycle, Boulder, Colorado), Jack Macy (SFEnvironmental, San Francisco), Robert Reed (Recology, San Francisco), Burr Tyler (GAIA), David Wimberly (Halifax), and Yuichiro

Hattori (Japan). In some cases they even went out and took new photos (David Wimberly, Yuichiro Hattori, and Kevin Hurley).

Every major change needs pioneers: people who dare to be different and who are prepared to reject old habits, however comfortable they may have become. These pioneers include people like Daniel Knapp, Mary Lou Van Deventer, Mark and Nancy Gorrell (Urban Ore, Berkeley, California); Robert Haley and Jack Macy (San Francisco); Ted Ward (Del Norte County, California); Bill Worrell (San Luis Obispo County, California); Eric Lombardi (Eco-Cycle, Boulder, Colorado); Buddy Boyd (Gibsons, British Columbia); Richard Anthony and Gary Liss (Zero Waste International Alliance); Neil Tangri, Monica Wilson, Cecilia Allen, Magdalena Donoso, Joan Marc Simon, and many others at GAIA; Sonia Dias (Brazil and Women in Informal Employment: Globalizing and Organizing); Jutta Gutberlet (Brazil and Canada); Bill Sheehan (Product Policy Institute); Neil Seldman and Brenda Platt (Institute for Local Self-Reliance); Jeffrey Morris (Sound Resource Management Group, Seattle); Annie Leonard (*The Story of Stuff*); Paul Hawken (author of *The Ecology of Commerce*); the late Tania Levy (author of *Garbage to Energy: The False Panacea*); Barry Friesen and David Wimberly (Nova Scotia); Helen Spiegelman (Vancouver, British Columbia); Luz Sabas, Von Hernandez, Sonia Mendoza, and Froilan Grate (the Philippines); Paul Martensson (Sweden); George Cheng (Taiwan); Gerry Gillespie and Graham Mannall (Australia); Warren Snow and Julie Dickinson (New Zealand); Katy Anderson (and colleagues at Cwm Harry, Wales); Jane Green (Friends of the Earth, Coventry, UK); Jim Henson (University of Coventry); Candida Brady and Jeremy Irons (director and host, respectively, of the movie *Trashed*); Mal Williams (Cylch, Wales); and many pioneers in the zero waste movement in Italy, including Rossano Ercolini and Mayor Giorgio del Ghingaro (Capannori); Pier Felice Ferri (and other members of Ambiente e Futuro, Lucca); Patrizia Lo Sciuto (Trapani, Sicily); Lucca Roggi (Pisa); Paulo Guanaccha (Catania, Sicily); Enzo Favoino (Scuola Agraria del Parco di Monza, near Milan); Roberto Cavallo (Piedmont, President of International Association Environemntal Communication, AICA); Professor Andrea Segre (University of Bologna); Riccardo Pensa (Fondazione Volontariato e Partecipazione); Concetta Mattia (ANPAS); Tomasso Sodano (vice mayor of Naples); Sergio Apollonio (leading opponent of the infamous Malagrotta landfill near Rome); and Pietro Angelini (Effecorta, Capannori); and over one hundred mayors in Italy whose communities have adopted zero waste (as of October 2012).

I am very grateful that nearly twenty of these pioneers have written guest essays for, or made other important contribution to, this book. So while only my name appears on the cover, there is no question that this has been a huge team effort.

PART ONE

Zero Waste Philosophy, Practicalities, Obstacles, and History

Historically we have had three answers to our trash woes: burn it, bury it, or cart it out to sea and dump it. It turns out none of these is really a good idea, one is now banned worldwide (ocean dumping), and none actually solves our problems. There is, though, a fourth way—zero waste—that's far better for both the local and global community.

The zero waste concept we're about to explore is relatively new. Even in the mid-1980s the idea would have been unthinkable in most regions around the world. But by 1995 the idea that we could cut consumption, recycle or reuse the waste we did produce, and prod manufacturers to design products with zero waste goals was beginning to fall on very fertile ground in places like California and Italy. From that moment, the zero waste ideal began to spread throughout the environmental movement and the more enlightened business community. It has taken hold in pioneering communities—both rich and poor, large and small—that have sought alternatives to landfills, incinerators, and a throwaway culture. Many of these communities have vastly reduced their waste, and many are looking to reduce it entirely by 2020.

For many of those not familiar with these pioneers and their successes, the zero waste goal may sound like pie in the sky. In reality, though, it can be approached in a series of just ten practical steps that are relatively simple, cost-effective, and politically acceptable. And, approached correctly, zero waste initiatives can put environment and economics on the same side, as they always can be with enough space and enough time.

To understand the goal, though, you first need to understand a bit about the philosophy behind it and the history of the movement. You also need to understand the role of incinerators in thwarting it. Even today, when incinerators are promoted as green alternatives because they are able to capture a little energy while destroying material resources, they set back communities that build them by at least twenty-five years in genuine efforts to move toward sustainability.

We don't have that kind of time to waste. So, the call to action begins.

The Big Picture

While I was writing this chapter, a bulletin from the Worldwatch Institute showed up in my in-box. The DC-based organization had just issued a report on the global state of solid municipal waste, and the news was not good. New research showed that the annual volume of that waste could double by 2025, thanks to growing prosperity and urbanization. Translation: Rather than producing 1.3 billion tons per year, as we do now, we could soon be producing 2.6 billion tons.[1]

The warning sounds gruesome, and indeed it is. But the future doesn't have to arrive smothered in trash. This doubling of global municipal waste production assumes a business-as-usual approach not only to waste management but also to all the other processes and decisions that make this waste in the first place. However, there is a different way of using and handling our resources. Unless we pursue this different approach, there is no way that we can edge our civilization toward a sustainable lifestyle on our finite planet.

Our current age is sleepwalking. Most of us living in Western (or Global North) societies have nearly everything our parents and grandparents ever dreamed of—except one thing, sustainability. We cannot expect our current consumption patterns to last very far into the future. We would need at least two planets to provide the world's population with the typical European consumption patterns and at least four planets if everyone consumed as much as the average American[2, 3] (see figure 1.1).

Simply put, we are living on this planet as if we had another one to go to. Overconsumption has been seen by some to be a disease of modern civilization. A word has even been coined for this social disease, *affluenza*—named after a popular documentary and book.[4, 5] Afflicted with consumerism, we see life and happiness as acquiring a series of objects; but study after study finds that, after a certain point, happiness does not grow alongside possessions.

Something has to change, and the best place to make that change is with waste, because everyone makes waste—every day. All the time we make waste we are part of a nonsustainable way of living on the planet, but given the right leadership we could all be part of an important first step toward sustainability. Achieving zero waste will also force us to look for other ways of finding happiness and require us to distinguish between what we desire (or others have persuaded us that we desire) and what we actually need to be happy.

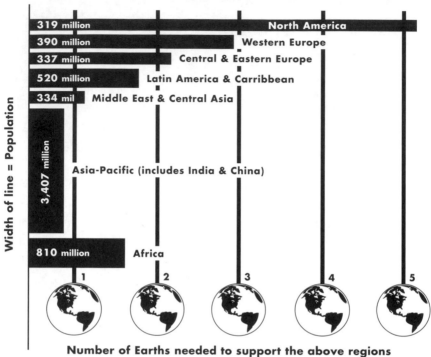

ECOLOGICAL FOOTPRINT BY REGION

Width of line = Population

- 319 million — North America
- 390 million — Western Europe
- 337 million — Central & Eastern Europe
- 520 million — Latin America & Carribbean
- 334 mil — Middle East & Central Asia
- 3,407 million — Asia-Pacific (includes India & China)
- 810 million — Africa

1   2   3   4   5

Number of Earths needed to support the above regions

FIGURE 1.1. Global footprints by country. *Illustration by Jack Cook*

Waste is the evidence that we are doing something wrong. Our task in the twenty-first century is not to find more and more sophisticated ways to destroy our material resources but to persuade industry and retailers to stop making products and using packaging that have to be destroyed.

The modern city has the cathedral of consumption at one end of town (the shopping mall) and a high-tech toilet (the incinerator) at the other. In between we run the throwaway society with many items in our hands for just a few seconds before they move from the "cathedral" to the "toilet"! (See figure 1.2.) Essentially this is a mindless process: Most people simply don't think about it. We have grown accustomed to running a throwaway society on a finite planet, but there are rational and practical solutions to waste and resource management that will move us in a sustainable—and less toxic—direction.

A Global Crisis: Waste Management in a Linear Society

Not only is the throwaway society presenting us with a local waste crisis, it is contributing to a global crisis. Global warming is only one manifestation of

Moving from the Cathedral of Consumption to the High-Tech Toilet

Shopping Mall

Municipal Incinerator

The Great Throwaway Society

Many items we may have in our hands for only seconds will end up moving quickly from the cathedral to the trash

SALE TODAY!

FIGURE 1.2. The "modern" city: Materials move very quickly from the cathedral of consumption (the shopping mall) at one end of town to the high-tech toilet (the trash incinerator) at the other. *Illustration by Jack Cook*

that crisis. A combination of overpopulation and overconsumption is using up the planet's resources at an ever-increasing rate, whether we are talking about fossil fuels, available clean water, arable land, rain forests, minerals, or fish. It is important to see what has caused this crisis and how a Zero Waste 2020 strategy can take an important step toward reversing the process.

Since the industrial revolution we have attempted to impose a linear society on a planet that functions in circles. Nature recycles everything; we do not. In four steps we convert virgin materials into waste. It starts with extraction of raw materials, which are often shipped halfway around the world. This is followed by three more steps: manufacture, consumption, and finally waste disposal. The more "developed" a society the quicker this transformation takes place.

Each step in this linear chain causes enormous impacts on the environment. Extraction from raw materials requires large quantities of energy and in turn produces huge quantities of solid waste, air pollution, water pollution, ecosystem damage, and massive quantities of carbon dioxide and other gases—which in turn lead to global warming. Most of these impacts are repeated again with the manufacture of products. Then transportation between each step entails further energy use and even more carbon dioxide production and even more global warming. Annie Leonard has produced an excellent video summarizing this linear society, called *The Story of Stuff.*[6]

In the end we have, of course, discarded materials; the way we handle that waste traditionally does nothing to mitigate the impacts of the linear society. Landfills bury the evidence and incinerators burn the evidence. This is true of incinerators no matter what fancy name is used to describe them—waste-to-energy plants, thermal valorization, gasification, pyrolysis, or plasma arc facilities.

In a nutshell, here are the contemporary options we have for dealing with waste—some that move us away from sustainability, and some that move us toward it.

> **Landfills.** When we bury discarded material in landfills, we are forced to go back to square one of the linear society, namely, the extraction of virgin materials (see figure 1.3). Not only does this approach not mitigate the impacts of the linear society, but it actually makes things worse. Any organic materials that are landfilled are going to be broken down by anaerobic processes to yield the gas methane, which molecule for molecule is about twenty-five times worse than carbon dioxide as far as global warming is concerned.
>
> **Incineration.** When we burn discarded materials in an incinerator we are again forced to go back to square one of the linear society. No impacts are mitigated. All that the incinerator industry can do is to claim that burning organic materials produces less global warming than landfills because *molecule for molecule* carbon dioxide from incineration produces less global warming than methane from landfills and is thus the lesser of two evils. But this comparison is simplistic and self-serving, as explained below. Nevertheless, at this very moment highly paid consultants and lobbyists are busily trying to persuade both state and federal governments to part with taxpayers' money to provide massive subsidies to help build incinerators that they claim can generate alternative energy and fight global warming.
>
> But as we'll discuss later, the rewards are far fewer than the consequences. The industry, wizened by its earlier struggles on the environmental front, has learned how to greenwash. Energy generation and the promise of less climate impact don't change the fact that when you burn resources they have to be replaced, which wastes more energy and creates more global warming than the marginal saving from reduced landfilling. Even the comparison between carbon dioxide production in incinerators versus methane production in landfills is simplistic. The methane in landfills is produced far more slowly than the carbon

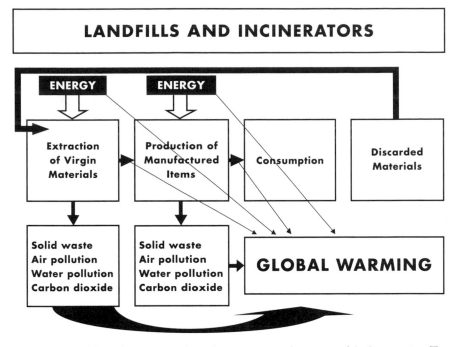

FIGURE 1.3. Landfills and incinerators do nothing to mitigate the impacts of the linear society. The arrows in this diagram indicate that when materials and objects are burned or buried society is forced to return to the beginning of the linear society (i.e., extraction of raw materials). *Illustration by Jack Cook*

dioxide in incinerators, which is essentially produced instantaneously. Also, synthetic materials like plastics are not broken down by anaerobic microbes in landfills to produce either methane or carbon dioxide—but when burned they do produce carbon dioxide.

Recycling Materials. Recycling discarded materials back to industry so that they can be used again in their manufacturing processes saves the impacts of the extraction of virgin materials and the energy costs of shipping them (see figure 1.4).

Reusing Objects. Reusing and repairing whole objects, rather than simply recycling their constituent materials, cuts out both the impacts of extraction of raw materials and the manufacturing processes.

Composting Organic Materials. Unless composted with organic waste materials from farms, compost is not a complete substitute for high-energy synthetic fertilizers, but it is a substitute for topsoil. Thus composting cuts out some of the impacts of the use of primary materials and high-energy fertilizer production (see figure 1.5).

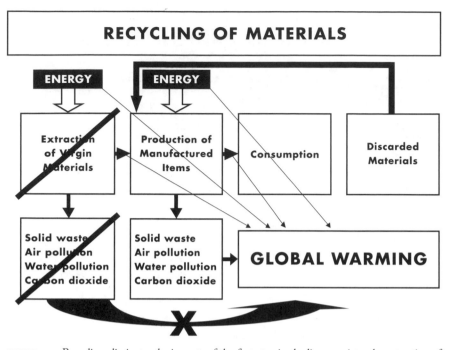

FIGURE 1.4. Recycling eliminates the impacts of the first step in the linear society: the extraction of virgin materials. *Illustration by Jack Cook*

However, composting does even more than this. By sequestering carbon in the form of cellulose, composting allows carbon to remain in the soil for many months or years before being converted to carbon dioxide or methane. Such cellulose in an incinerator is instantaneously converted to carbon dioxide. Composting also returns nutrients to the soil and helps to retain moisture and fight erosion.

Avoiding Waste and Reducing Consumption. Consumption is the ultimate driver of this whole linear process, and it is stimulated by advertising, particularly via today's number-one human pastime: watching TV. Overadvertising produces overconsumption. In the United States, every seven minutes we are told that we need to buy something. We are told that we are hungry, thirsty, too fat, too sick, sexually frustrated—and need a new car! By the time a high school student leaves school in the United States, he or she will have watched over 350,000 TV commercials.[7] From a very early age our children are being programmed for an overconsuming lifestyle.

Yet if we find ways of avoiding the consumption of unnecessary items, we can cut out many of the impacts of the linear society. A lot of this

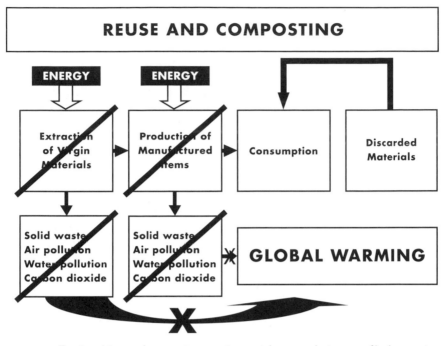

FIGURE 1.5. Reusing objects and composting organic material cuts out the impacts of both extracting raw materials and the manufacturing processes. The arrow in the figure indicates the flow of reusable objects and compost back to the consumer. *Illustration by Jack Cook*

overconsumption is driven by ever-changing manufactured "fashions." As Oscar Wilde said, "What is fashion? . . . It is usually a form of ugliness so intolerable that we have to alter it every six months." On a more serious level, Mahatma Gandhi said that, "There is enough in the world for everyone's need, but not enough for everyone's greed."

With China, India, Brazil, and other countries with huge populations attempting to catch up with the Global North's consumption patterns, the stresses on finite resources and global warming are destined to become far, far worse than anything we have seen to date. It is time that we in North America, Europe, and Japan set a better example. Simply put: A sustainable society must be a zero waste society.

What Is the Zero Waste Strategy?

The zero waste strategy says no to incinerators, no to megalandfills, no to the throwaway society, and yes to a sustainable society. While it may sound like an idealistic goal, we can put it into a realistic time frame. We do not expect to reach zero waste next year, but we can anticipate that some communities could be very close to eliminating most of their waste by 2020.

People will quibble about how close we can get to producing zero waste, but the point is that by aiming for zero we make our intentions very clear—and we're more likely to get closer to that goal than if we set ourselves a lesser target. Put another way, those in the zero waste movement pose this question: Bearing in mind the needs of future generations, how much waste do you think is acceptable? In short, how much waste are you *for*?

California-based zero waste advocate Gary Liss once explained the zero waste goal this way: "Every day, more than 100 million citizens do the right thing . . . they recycle. Now it is time to set our sights higher and start planning for the end to wasting resources and to our reliance on landfills, incinerators and other waste facilities. Zero waste is a policy, a path, a direction, a target; it's a process, a way of thinking, a vision."

For me, more than anything else zero waste is a new direction. We have to move from the back-end of waste disposal to the front-end of resource management, better industrial design, and a postconsumerist way of life. In both industry and our daily lives we have to get waste out of the system. As Mary Lou Van Deventer and her husband, Daniel Knapp (two entrepreneurs that have turned waste reduction into a business), say, "Discarded materials are not waste until they are wasted. Waste is a verb, not a noun."

In 2004 the Zero Waste International Alliance (ZWIA) established the only peer-reviewed definition of zero waste to date:

> Zero Waste is a goal that is ethical, economical, efficient and visionary, to guide people in changing their lifestyles and practices to emulate sustainable natural cycles, where all discarded materials are designed to become resources for others to use.
>
> Zero Waste means designing and managing products and processes to systematically avoid and eliminate the volume and toxicity of waste and materials, conserve and recover all resources, and not burn or bury them.
>
> Implementing Zero Waste will eliminate all discharges to land, water or air that are a threat to planetary, human, animal or plant health.

The Four Rs

In my talks to communities fighting incinerators I have had some fun at the expense of high-paid consultants who push incinerators and landfills. I say

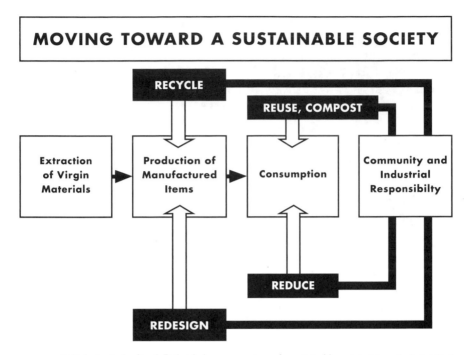

FIGURE 1.6. Redesign is the fourth R that helps us move toward a sustainable society. *Illustration by Jack Cook*

that they are thinking with the wrong end of their bodies. They are what I call "back-end thinkers." The back-end thinker comes home and finds the bathtub is overflowing so he (and it usually is a *he*) quickly grabs a cup to empty it. That doesn't work, so he grabs a bucket, then a foot pump, and then an electric pump—all in a vain effort to empty the bathtub before it damages the floor. Then his wife comes home and switches off the tap. She is a "front-end thinker"!

Many front-end thinkers have long embraced what have come to be known as the three Rs (reduce, reuse, recycle)—but when it comes to waste, which is ultimately a design problem, we need to add another R to the lexicon: *redesign* (see figure 1.6). In fact, one of the first Italians to talk about "zero waste" was the greatest designer of all time—Leonardo da Vinci. In his writings he said that there was no such thing as waste. Like nature, each producer's waste product should be another producer's starting material.

Redesign is important in making the fundamental transition from the back-end to the front-end of the waste-resource problem as well as grabbling with the larger issue of sustainability. To make this transition we need all sectors of society to accept some responsibility. We need individual responsibility, community responsibility, industrial responsibility, professional

ZERO WASTE—a New Direction

Not just
Waste Management

Community Responsibility
(the back-end)

NO—to a   YES—to a
Throwaway   Sustainable
Society   Society

NO   NO
Incineration   Landfills

ZERO WASTE BY 2020

Responsible Political Leadership
(joining Community & Industry)

Industrual Responsibility
(the front-end)

and participation from
All Citizens

Sustainable Industrial Design
MOVING TOWARD THE FRONT-END

FIGURE 1.7. Zero waste is a new direction. We need to move from the back-end of waste disposal to the front-end of better industrial design. *Illustration by Jack Cook*

responsibility, and political responsibility. We need community responsibility at the back-end of the problem, industrial responsibility at the front-end, and we need responsible political leadership to bring these two together. Moreover, because this issue is too important to leave to the narrow focus of waste disposal experts, we need every sector of the society and economy involved. We need to integrate zero waste planning with farmers, doctors, artists, educators, communicators, philosophers, scientists, engineers, economists, environmentalists, industrial workers, architects, community developers, social activists, and children! In short, we need everyone involved in this massive effort (see figure 1.7).

But even as a massive effort is contemplated, it's critical to remember that communities recycle—not countries. While national, state, and regional governments can pass helpful legislation—and sadly unhelpful legislation (for example, giant subsidies to incinerators)—the key issue is what your local community can do. The best way a new zero waste advocate can take action is by finding out what other communities of similar size and demography have been able to achieve, then choose the best model and try to copy and improve on it. That's why you'll find, in Part 2 of this book, a guidepost to the best and most replicable zero waste projects the world over.

When we are looking at what has contributed to the best examples of zero waste to date, I would say that the single most important factor is the willingness of decision makers to work with citizens (and corporate decision makers with their employees) to find solutions we can all live with. Citizens are certainly going to be the enemy when decision makers opt for resource-destroying and polluting landfills and incinerators, but they are the key allies when decision makers seek genuine solutions that move us in a sustainable direction. Similarly, when corporate leaders add two more Ps to *profit* (*people* and the *planet*), they will get huge support from their workers. Such solutions are not only better for the local community and the corporation, but better for the planet as whole. We all need a larger meaning to our work and our lives. Money is not enough.

Summing It Up with the Four Cs

We have the three Ps of responsible business (people, profit, planet) and the three (or, we should say, four) Rs of environmental stewardship, and now it's time to introduce the four Cs of sustainable progress—critical concepts to keep in mind as we move from a linear, throwaway society to a circular and sustainable society.

Common Sense. The ten steps to zero waste I outline in chapter 2 are little more than common sense in action. Many people who put their minds to it could have worked out the same system for themselves—and many have done so. It doesn't require a PhD. Once one recognizes that waste is made by mixing discarded materials together, the rest of the steps follow in a fairly simple and logical sequence.

Community. With little evidence that our national leaders are willing to get together and take the tough steps needed to move our societies toward sustainability, we have to save our planet one community at a time. Part of that reclamation is taking charge of our own waste (discarded resources) streams. We need good local leaders who instigate the simple systems required and citizens who support that leadership by cooperating with the steps needed.

Creativity. Creativity at the front-end can save millions at the back-end. Think of the best food packaging in the world—the ice cream cone. We eat the ice cream and then we eat the package! Think of the reusable glass bottle that can go around up to forty times before it is recycled. Think of reusing the old lumber in buildings to make beautiful furniture. Think of the volunteers who collect food from supermarkets before the due date and get it to homeless shelters. Think of those who train to be master composters and then show their neighbors how to compost their food scraps in their own backyards. Think of the people who take scraps of derelict land in cities and convert them into gardens and use up composted food scraps in the process. There are so many wonderfully creative ideas happening and waiting to happen in waste resource management once we move away from careless and mindless disposal methods like landfills and incinerators.

Children. If we can get our children involved in this issue early—and excited about it—many more creative ideas will flow into the system. Moreover, we have to give our children hope about the future. I am afraid it is far too easy to depress our children by what we are doing to our beautiful planet. It is also too easy to deny what we are doing and keep it hidden from their eyes. Neither approach will work. We need to be honest with ourselves and with our children about the devastating problems that we have created. But we need to couple this honesty with an action plan in which our children have a key role to play. The reward for any child, or student or citizen, for getting involved in the zero waste effort is that by fighting the greatest challenge that we have been handed since the industrial revolution they will seldom be bored. Angry, yes. Frustrated, yes. But bored, no. Most importantly, engaging in creative problem solving for a sustainable future gives children the chance to find genuine meaning. Finally, we need our children involved because it is their future. It is they even more than us who are going to need to create new ways of relating to the world.

We have all grown accustomed to the task of reshaping the world to fit our needs. Now we have to reshape our needs to fit the demands of the world. In other words we need to find ways to balance the demands of an ever-increasing population with what it will take to survive within a finite biosphere.

The good news is that there is a very down-to-earth, practical way to begin the journey to zero waste: It begins with the ten fingers on all our hands, and culminates with the brightest and most creative minds in our society.

Ten Steps Toward a Zero Waste Community

It may appear that zero waste is an idealistic goal, but we can approach this target with simple, practical, cost-effective, and politically acceptable steps in mind. In fact, as you'll see in Part 2, many cities and towns around the world are taking these steps to get closer to—and hopefully eventually achieve—zero waste.

You'll note that many of these steps stress community—another reminder that getting closer and closer to zero waste involves building both local and global communities in order to move toward the greater goal of sustainability. We'll see how cities and towns around the world are implementing these steps in Part 2, but for now the keys to moving these ten steps forward lie in both community and industrial responsibility.

Step 1: Source Separation

Waste is made by mixing discarded items. Waste is unmade (or rather not made in the first place) by keeping discarded materials separated into a few simple categories. While many communities in the United States recycle, the types of items collected at curbs are usually limited to three or four categories (see Step 2 below). However, in places where citizens deliver their discards (materials they no longer need) directly to drop-off recycling centers, the number of categories can be much higher. In one town in Japan, citizens separate into thirty-four different categories (see chapter 11). Daniel Knapp of Urban Ore, a leading recycling venture in Berkeley, California, believes all discarded materials can be divided into twelve master categories, and he designs resource recovery parks in ways to receive and handle these categories[1] (see chapter 15). It all boils

The Ten Steps at a Glance

1. Source separation
2. Door-to-door collection systems
3. Composting
4. Recycling
5. Reuse, repair, and deconstruction
6. Waste reduction initiatives
7. Economic incentives
8. Residual separation and research facilities
9. Better industrial design
10. Interim landfills

down to the simple fact that the more categories a community offers, the more waste its residents can recycle and the less material is wasted in processing.

Step 2: Door-to-Door Collection Systems

Door-to-door collection systems in towns that are serious about zero waste typically involve three or four color-coded containers or bags. Collection rates vary, as does the degree of curbside waste separation. San Francisco, for instance, does a once-a-week collection of three containers (see chapter 6).[2] Several cities in Italy and Spain use four or more containers, and some communities collect specific materials (using the same collection vehicles) on different days of the week to allow for greater source separation. Communities at the leading edge of door-to-door collection have been able to allow citizens to recycle over 80 percent of their discards, and some smaller communities over 90 percent.

Basically, in the more comprehensive door-to-door systems one container is used for kitchen waste, one or more for recyclables (an extra division can be made here between paper products and bottles and cans), and a third for the residual fraction. Being produced seasonally, garden debris is usually picked up less frequently. Some communities collect the kitchen and garden debris together, but this leads to the use of much larger containers—which in turn can influence the size and expense of the vehicles used for pickup. Most communities in North America, though, still lack door-to-door pickup of organics and kitchen waste.

Step 3: Composting

In my view composting is more important than recycling. Kitchen waste and other organic matter not only causes foul odors when mixed waste is left around in cities, it also generates methane and leachate in landfills. But perhaps the most important reason that we need to collect *clean* organic waste is because it is needed by farmers to replenish their soils of depleted nutrients and, especially in warmer climates, help fight erosion by holding onto moisture. Not only does the conversion of organic matter to compost avoid the global warming gases involved in the production of synthetic fertilizers and topsoils, but it also delays the release of global warming gases from the waste materials themselves by sequestering the carbon in wood and other cellulosic fibers in the final product. With incineration, the conversion of cellulose and other organic material to carbon dioxide is instantaneous; the cellulose left in compost can last in the soil for many months to many years. Moreover, by

FIGURE 2.1. A farmer holds up a mustard plant used as a cover plant in the rich soil between the rows of vines in one of the nearly two hundred vineyards that use the compost produced from the organic material collected in the city of San Francisco. The long roots on this plant represent captured carbon, which will be held in the soil for many months or even years. *Photograph by Larry Strong, courtesy of Recology*

stimulating plant growth, this in turn leads to more absorption of carbon dioxide from the air (see figure 2.1).

In San Francisco the kitchen and other organic waste is sent to a large composting plant located approximately one hundred kilometers from the city. The site is surrounded by farmland, and local farmers use the compost to produce fruit, vegetables, and wine—which is sent back to San Francisco (see chapter 6).[3, 4]

According to Robert Reed, a spokesperson for Recology, the company that runs the composting plant for San Francisco, the city has composted 1.2 million tons through its green bin program since it began in 1996. By so doing, Reed says, "San Franciscans have achieved a total CO_2 equivalent benefit (methane avoided and carbon sequestered) of at least 640,000 metric tons. That is equal to offsetting all emissions from all vehicles crossing the Bay Bridge for three and a half years."[5] (See figure 2.2.)

Instead of exporting their mixed waste to landfills and incinerators located in rural areas, which causes so much intense opposition from citizens and farmers, municipal decision makers should work with farmers to coproduce a compost product from which everyone can benefit. Even so, municipal composting facilities need to be located carefully because of the odor problem—not necessarily a huge problem in rural settings but highly problematic in or near cities.

Farmers do not want low-grade material, and so the key to a municipal composting program's success lies in the ability of cities to organize their citizens to separate their organic discards from plastic, glass, and other contaminating materials. In this respect, the city workers who pick up this material can be very important players in the education process as well as the kitchen staff in hotels and restaurants.

Many towns where householders have more space have taken a simple preliminary step before building a centralized composting facility. They encourage as many of their citizens as possible to compost their own kitchen and yard waste in backyard compost bins or vermiculture boxes (worm bins). Some provide the composting or vermiculture kits either for free or at reduced

FIGURE 2.2. Thirteen years of composting in San Francisco have sequestered enough carbon and prevented enough methane releases to offset three and a half years' worth of emissions from Bay Bridge traffic. *Photograph by Larry Strong, courtesy of Recology*

FIGURE 2.3. *From top:* a farmer inspects the compost produced from San Francisco's organic discards; compost being spread on the farm owned by Recology; and a farmer proudly showing the tomatoes he has grown using this compost. *Photographs by Larry Strong, courtesy of Recology*

cost. Communities are sometimes aided by nongovernmental organizations that train volunteers to become "master composters." These show people how to get started and troubleshoot problems as they arise; such a program is run by the UK group Garden Organic.[6]

Zurich, Switzerland, which has a very dense housing situation, has encouraged "community composting." In this program a number of households (ranging from three to two hundred) share the responsibility of running a simple compost system. These do not occupy a large area and can be located in city parks or in the space between high-rise buildings. Currently the city boasts over one thousand community-composting plots, which in total are taking care of about half of the city's household organic waste. According to Thomas Waldmeier, who pioneered this program, the best thing about it has been its social impact: "It helps people fight the anonymity of living in a big city," he said. "People meet over the compost pile!"[7]

Every town or city needs to design a program that works for its needs, but they all need to assess the same factors for handling food waste and establishing priorities. Here's how I would prioritize those factors:

1. Feeding humans using in-time marketing of food items close to their sell-by dates from supermarkets, as well as unused food from restaurant kitchens
2. Feeding animals
3. Home composting or vermiculture
4. Community composting or vermiculture
5. Small-scale in-vessel composting systems in urban areas
6. Co-composting with agricultural waste in rural areas
7. Centralized composting or anaerobic digestion in rural areas

Step 4: Recycling

In larger communities, recyclable materials are destined to go to material recovery facilities (MRFs), of which there are hundreds of successful examples around the world. Their function is to separate the paper, cardboard, glass, metal, and plastic and prepare them to meet the specifications of the industries using these secondary materials to manufacture new products.

Some of these plants are built to handle a single stream of mixed recyclables; others deal with two streams—paper products in one stream and glass, metal, and plastic (like cans, bottles, and food containers) in the other. As far as the industries that use these secondary materials are concerned, the simple rule is that they want three things: quality, quantity, and regularity.

FIGURE 2.4. Typical materials recovered at the Pier 96 MRF in San Francisco, including various grades of plastic, paper, and cardboard. The value of these materials continues to rise. *Photograph by Larry Strong, courtesy of Recology*

These plants are best built in large cities, which can provide the large labor force needed and are also usually located close to the industries that can use the secondary materials or have transportation hubs for delivery elsewhere. This sets up an ideal partnership between urban and rural areas. The cities should export their clean, source-separated organics to the rural areas for composting, and the rural areas should transfer their recyclables to the cities so that these materials can be sent back to industry.

Sadly, in many large cities in the United States the recyclables are being shipped to China instead of to local industries. However, in Nova Scotia (a Canadian province of about 900,000 inhabitants) nearly all the recyclables are used in the province's own industries (see chapter 9). Their program has created approximately one thousand jobs in the collection and processing of the discarded materials (see figure 2.4) and another two thousand jobs in the industries using these secondary materials.[8]

Some zero waste advocates would prefer to see recyclables being dropped off or delivered to zero waste ecoparks with early separation into more categories. These views get a greater airing in chapters 15 to 17.

Step 5: Reuse, Repair, and Deconstruction

There are many successful examples of reuse and repair centers running as either for-profit or not-for-profit entities. An example of the former is Berkeley's

Urban Ore, which has been running for more than thirty years. It receives appliances, furniture, and other items from homeowners and also materials and items obtained from companies that specialize in the deconstruction (as opposed to demolition) of old buildings. These building and renovation materials are a very profitable part of their business. Timber, bricks, bathroom fittings, doors, and windows have a high resale value. More and more builders are dropping off their recovered materials at Urban Ore while they are picking up reusable items for new projects.[9]

Urban Ore currently grosses nearly $3 million annually and has about thirty full-time employees, who are well paid and have a good benefit system. Some of them have worked for the company for more than twenty years. The company accepts anything reusable and lays out the household goods like a department store.

Building deconstruction and renovation is a particularly rich source of reusable items at Urban Ore. Deconstruction takes longer than demolition, but it yields more employment and recovers valuable materials. At Urban Ore some of these materials are stored and displayed in an outside yard, and smaller and more valuable items are on view inside (see figure 2.5).

Urban Ore will pay for valuable items, but more often than not people are only too happy to see their secondhand appliances and furniture used again and not simply crushed and sent to a landfill or burned in an incinerator.

Another good example of a reuse operation is ReSOURCE (formerly ReCycle North) located in Burlington and Williston, Vermont. ReSOURCE specializes in fighting poverty both by providing free goods for people in difficult financial situations and with job training. People, literally taken off the street, are trained to repair goods in one of five different categories: large appliances, small appliances, electrical goods, electronics, or computers. After nine months the successful trainees are issued with certificates and given help to secure full-time employment.[10]

On its website ReSOURCE gives a summary of its history and its mission:

> In 1991 in Burlington, Vermont, ReCycle North began an innovative program of repairing and reselling household items that otherwise would have gone to the dump. Out of that founding vision, more than 750 people have received job training and skills essential to gainful employment, 10,000 low-income people have received needed household goods and building materials, more than 10,000 tons of materials have been kept from the landfill,

FIGURE 2.5. The building materials area inside the Urban Ore building. *Photograph by Patrizia Lo Sciuto*

and 50 people now have secure employment through income earned largely from this social enterprise. ReSOURCE's success to date has proven its sustainability and is replicating this model of environmental sustainability, educational training, and economic opportunity in central Vermont.[11]

ReSOURCE also runs a deconstruction operation, and some of the recovered wood is used to make furniture (see figures 2.6 and 2.7).

FIGURE 2.6. Jeremy Smith, an employee of ReSOURCE's Waste Not furniture division, applies the finishing touches to a piece of furniture fabricated from lumber recovered from an old building. *Photograph by Kevin Hurley, PolarisMediaWorks.org*

Like the Salvation Army and Goodwill Industries International, ReSOURCE is also able to attract higher-priced items by allowing people to treat the objects that they give as tax-deductible donations.[12]

Both Urban Ore and ReSOURCE offer something more than good bargains. People enjoy visiting these facilities, and thus they have become centers for community interaction and activity. Having places like this where people can meet and interact is invaluable; like the community compost piles in Zurich, they help to fight the anonymity of living in a big city.

In an ideal world such facilities would be located throughout the city as a way of stimulating community development—a conscious effort to remake the "village" within the city. To this end some space should be set aside at such facilities for community meetings and entertainment—as in Gothenburg, Sweden, where a very creative reuse park is being used to stimulate community interest and involvement (see chapter 10).

These operations work so well because reusable items are valuable. Recyclable materials are high volume and low value, while reusable objects are low volume and high value. In Los Angeles it is estimated that only 2 percent of the domestic waste stream is reusable, but it is worth well over a third of the total value of all the discarded material.[13] In the United Kingdom in 2010,

FIGURE 2.7. Some of the furniture and other products made from old lumber by the Waste Not furniture division of ReSOURCE, located in Burlington and Williston, Vermont. *Photograph by Kevin Hurley, PolarisMediaWorks.org*

the city of London spent approximately $12 million to set up nine reuse and repair centers scattered around the city.[14] They have a special focus on office furniture and building materials.

Reuse and repair facilities thrive financially because all kinds of people enjoy searching for a bargain. In the United Kingdom, every High Street has thrift stores that support various charities. These are well equipped to handle smaller items like books, CDs, toys, and clothing, but a community reuse and repair center is better designed to handle larger items like appliances, furniture, and building materials while providing facilities for repair and training. In the future, space can be made for part of the zero waste research operation (see Step 8 below). The repair of products provides the most appropriate moment to investigate their design and perhaps recommend redesign. Hopefully, such research will provide a push for the elimination of built-in obsolescence.

In addition to the recovery and resale of objects, a reuse and repair center can be used as a collection center for hard-to-recycle items and those containing toxic materials—like old paint, batteries, fluorescent bulbs, and solvents. Either the manufacturers who use these facilities (to collect items covered by extended producer responsibility [EPR] programs) or municipal authorities (anxious to keep toxics out of their landfills) should anticipate paying fees to

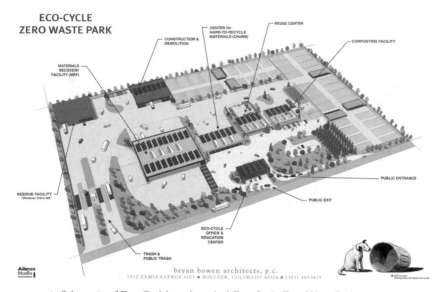

FIGURE 2.8. Schematic of Eco-Cycle's trademarked Eco-Cycle Zero Waste Park. *Photograph courtesy of Eco-Cycle, Boulder, Colorado*

reuse and repair centers providing this service. Guest author Daniel Knapp discusses this further in chapter 20.

Three zero waste leaders, Knapp, Buddy Boyd, and Eric Lombardi, have argued for reuse and repair centers to be expanded into ecoparks dedicated to handling the whole of the source-separated discard stream from both households and commerce. Such one-stop shopping could prove to be more effective in combating incinerator companies who promise to take the whole of the residual fraction at one facility. Knapp—with his late friend and colleague architect Mark Gorell—designed many of these ecoparks and shared them with decision makers in places as far away as Hawaii and Australia.

Canberra, Australia, prior to the demise of its No Waste by 2010 program,[15] was on course to build such a "resource recovery park."[16]

The closest in fulfilling the ecopark dream is Lombardi, who runs the very impressive recycling operation Eco-Cycle in Boulder, Colorado (see chapter 6). Using his concrete experiences, he is offering to design and bring together individual manufactures and operators for the building and running of such a park, as illustrated in figure 2.8.

Step 6: Waste Reduction Initiatives

The separation and collection of clean organics, marketable recyclables, and reusable objects takes us closer to a sustainable future—but the residuals do

The First Five Steps

In October of 2012, San Francisco (population 850,000) reported that by using a combination of these first five steps—source separation; door-to-door collection systems; composting; recycling; and reuse, repair, and deconstruction—the city had reached a diversion rate of 80 percent from landfills (it has no incinerator) en route to their goal of zero waste by 2020 (see chapter 6).[17] It should be noted that an 80 percent diversion is higher than the city would have achieved if it had built an incinerator in 1985 as was proposed at the time. An incinerator only achieves a reduction of 75 percent by weight as 25 percent is left as ash that nobody wants (see chapter 4). Currently, San Francisco is only sending 20 percent of residuals to landfill. So San Francisco has demonstrated to the industrialized world that it is totally unnecessary to go through the hugely expensive and problematic process of building an incinerator.

not. The residuals are our failures. They are failures of either poor manufacturing design or unwise purchasing habits. Before we deal directly with these failures (see Steps 8–10), we need to examine other things we can do to minimize their production.

There are many unnecessary items—especially packaging—that have entered our modern lives. As these pile up in landfills, more and more governments and private enterprises are taking steps to reduce their use and production.

In Ireland, the government introduced a fifteen-cent tax on each plastic shopping bag used in shopping malls. Much to the surprise of everyone, within one year this measure reduced the use of these bags by 92 percent. The tax netted 12 million euros to support other recycling initiatives.[18]

In Australia, many towns have banned the use of plastic shopping bags completely. As a result of these measures more and more people are being persuaded to use reusable bags made of cotton or other durable materials. These have the advantage of displaying the slogans and messages of sponsoring organizations. Bans on plastic shopping bags are spreading throughout the world,[19] and even large cities like San Francisco have introduced them.[20]

In Italy, some supermarkets have dispensing systems that allow customers to reuse their own containers for various liquid items like water, milk, wine, shampoo, and detergents. Other dispensing systems are being used for solid items like grains and cereals. In 2009, a new food store (Effecorta) was opened in Capannori, Tuscany, that uses dispensing systems for sixty different liquids and sixty different solids. Customers bring in their own refillable bottles or containers. Adding to the sustainability of this operation is the fact that over 90 percent of the food and beverage items sold are produced within seventy kilometers of the store. In the case of wine sold in the store, the place of origin is indicated on a map so customers can see where it is produced (see figure 2.9).[21]

Many older citizens remember a time when their beer and soft drinks came in reusable glass bottles with a refundable deposit on them. More and more of these have been replaced—supposedly for our convenience—by recyclable or disposable bottles. However, in one major jurisdiction—the province of Ontario in Canada—this has not happened for beer bottles. For over sixty years, the beer industry there has used refillable glass bottles. Today, according to the director of the Beer Store (which has a monopoly on beer sales in Ontario), over 98 percent of these bottles are recovered. Each bottle goes around eighteen times, and there is a calculated lifetime savings of thirteen cents per bottle compared to disposable glass bottles. More than two thousand jobs are involved in collecting and cleaning these bottles. Most importantly, this whole operation is conducted by the beer industry itself; there is no cost whatsoever to Ontario communities. Municipalities do not have to collect the bottles nor attempt to landfill, burn, or recycle them. The beer industry has internalized its packaging costs. So here we have essentially a zero waste operation—and it has been going for over sixty years! It is cost-effective, simple, practical, and certainly politically very acceptable.[22]

If we put our minds to it there are many other places where we can return to older and more sustainable habits and activities. One place is the cafeterias in schools, offices, universities, and other institutions. We need to get rid of all the disposable plastic items and go back to washing up ceramic cups and dishes, glass cups, and stainless steel cutlery (see figure 2.10). Here is where local governments could help by providing some of the capital for such conversions (washing machines), thereby conserving finite resources as well as dramatically reducing the costs of disposing of the waste in landfills or incinerators—as well as creating more local jobs.

Another old-fashioned way we can reduce waste, and save money doing it, is by encouraging more and more parents to use reusable rather than disposable diapers. It costs an American family between one and two thousand dollars to put a baby through disposable diapers.[23] Easy-to-use reusable diapers (with Velcro strips or snap buttons for easy replacement) and cellulose-based absorbent material, even at forty dollars or more, per each reusable diaper saves a great deal of money over a "babyhood" of disposable diapers.

FIGURE 2.9. Eight different locally produced wines can be sampled at Effecorta, a store in Capannori, Tuscany, which provides sixty dispensing systems for a variety of liquids. *Photograph courtesy of Pier Felice Ferri*

FIGURE 2.10. Students at a primary school in Capannorri, Italy, use glass, china, and stainless steel utensils and cutlery instead of disposable plastic items. *Photograph courtesy of Pier Felice Ferri*

Of course these are just a handful of examples from a huge, diverse residual waste stream. But the point is that a little creativity at the front-end can save millions at the back-end.

Step 7: Economic Incentives

One example of a powerful incentive to reduce the residual fraction is to use a pay-as-you-throw (PAYT) system. The idea here is to encourage citizens to maximize the diversion possibilities by penalizing the production of residuals. Typically, recyclables and compostables are picked up for free or at a flat rate (sometimes absorbed in local taxes), but an extra charge is applied to the residuals. The more you make the more you pay. This can be done in several ways: In some communities the residuals are weighed; in others you can choose the size of the container for the residual fraction but for the larger container you pay more (as in Seattle, Washington); in still others, stickers are purchased to affix to each bag placed on the curb; or special plastic bags might have to be purchased. This one simple fiscal step has led to significant reductions in many jurisdictions.[24, 25] For example, Villafranca d'Asti (with a population of three thousand) in the province of Piedmont, Italy, was able to go from a diversion rate of 70 percent to 85 percent by introducing such a PAYT system.[26]

However, such initiatives are not always popular. People argue that they are already paying for their waste to be picked up (in their local taxes), so they do not feel that they should be charged any more for the residual fraction. A response to this perception problem is to change the strategy slightly by making it into a "*save* as you throw" system. In this the council determines the expected or acceptable amount of residuals for a particular family size; if citizens put out less than this, they receive a rebate on their local tax bill.

With a combination of waste reduction initiatives and economic incentives communities can drastically reduce the residual fraction, but for the time being we will still need to find ways of dealing with the residual fraction remaining after Steps 1–7. This brings us to Step 8, the most important step if we wish to get close to zero waste by 2020.

Step 8: Residual Separation and Research Facilities

How the residual fraction is handled is the key difference between waste disposal (landfills and incinerators) and the zero waste strategy. The former attempts to make the residuals disappear, the latter sets out to make them very visible. We need these residuals made very visible if we are to move toward a sustainable society. We need to study and correct our mistakes.

Thus in the zero waste strategy the residuals need to be sent to a *residual separation and zero waste research facility* and not directly to a landfill. We will discuss this facility in two parts: the residual separation facility and the zero waste research center.

Residual Separation Facility

Figure 2.11 shows the Otter Lake Landfill, which is located a few miles outside Halifax, Nova Scotia. The site includes the residual screening facility, which has been built immediately in front of this landfill. Here the key concept is that the residual waste must be screened, sorted, and processed before any material enters the landfill.[27]

On arrival the bags of residual waste are opened and the contents tipped onto conveyor belts. Magnets pull out items made of steel or iron. The rest is passed along the conveyor belts to well-protected and trained personnel who pull out bulky items, more valuable recyclables (such as beverage containers on which there is a returnable deposit and office paper), and toxics.

The dirty organic fraction—for example, kitty litter, diapers, or unseparated food scraps—reaches the end of the conveyor belts untouched. This is then shredded and biologically stabilized either by a second composting operation (in the Nova Scotia example) or an anaerobic digestion system in other mechanical and biological stabilization facilities. The point of this process is not to produce an organic product for sale (it is contaminated) but rather to

FIGURE 2.11. The residual screening facility in front of the Otter Lake Landfill (at the top of the photo) screens, sorts, and processes waste before it can enter the landfill—making for a smaller, safer, and more politically acceptable landfill. *Photograph courtesy of Mirror Nova Scotia*

ensure that much of the organic degradation occurs above ground in a controllable fashion before it takes place underground in an uncontrollable fashion.

Nova Scotia's resulting landfills are far smaller and less problematic than the raw waste landfills that preceded them or the ash landfills needed by incinerators. Nova Scotia has pioneered this approach with some success. The overall Nova Scotia program, for which the residual separation facility is the linchpin, was designed by citizens based on consultant reports provided to them by the city of Halifax.[28, 29] More details of the program and photographs can be found in chapter 9.

Even without the incorporation of the zero waste research center, careful environmental comparisons indicate that these kinds of residual screening and biological stabilization facilities manage residual waste far better than any other disposal method. A recent report commissioned by Eco-Cycle compared eight options for handling residual waste and analyzing their impact on climate change, acidification, eutrophication (the result of too many nutrients in a water body), respiratory diseases, noncancers, cancers, and ecotoxicity. Those options included burying the residuals in landfills with various landfill-gas-to-energy scenarios; burning the residuals in waste-to-energy operations, followed by landfilling the ash, as practiced today by leading waste-to-energy companies; and mechanical recovery and biological treatment followed by landfilling, with two different assumptions for the recovery of recyclables and two different assumptions for gas collection efficiencies.

The results of this study showed that mechanical recovery and biological treatment followed by landfill "had the lowest overall environmental and human health impacts of all the disposal technologies."[30]

The Zero Waste Research Center

We can refine our waste handling procedures to yield smaller and smaller amounts of waste by removing anything recyclable, reusable, or toxic from the waste stream, but we need to do more in a zero waste program than simply landfilling the remaining residuals. We need to carefully observe and study them. This gives us our first opportunity to integrate zero waste with the higher educational system—a critical step if we want to use the zero waste strategy as a stepping-stone to sustainability.

Once communities have residual screening facilities, we need to add research centers. Ideally, these centers would be run by a local university or technical college. In this research center professors and students with various interests in a sustainable future (e.g., industrial design, clean production, ethical advertising, urban and community development, economics, environmental management, and global degradation) could study the nonsustainable mistakes of today's society and propose alternatives. For those interested in sustainability, the zero waste research center is an ideal laboratory.

Some of the research activities and actions at research centers could include the following:

- monitoring, evaluating, and improving the whole program by maximizing the capture rate of reusables, recyclables, and clean compostables in the door-to-door collection systems as well as the reuse and repair centers;
- preparing a compilation of best practices for waste reduction and avoidance strategies from around the region, the country, and the world and sharing these with decision makers and local businesses;
- developing some local uses for some materials (for example, shredded newspaper can be used for cattle bedding, building insulation, or in the production of papier-mâché as a substitute for Styrofoam molds for the transportation of fragile goods);
- developing alternatives to completely avoid or reduce the use of toxics in products (e.g., batteries, paint, solvents);
- recommending better industrial designs to industry on packaging and products.

In my view, the zero waste research facility is the key step in moving toward zero waste and sustainability. It is the *brain* of the system. Like a doctor, it diagnoses the maladies and attempts to provide the therapies. It monitors the system and provides the discipline we need to keep moving forward. Like nature it provides feedback mechanisms on bad industrial design (see figure 2.12). We have a waste problem because for so long we have refused to introduce feedback mechanisms, relying instead on landfills and incinerators to get rid of the things we can't immediately reuse, recycle, or compost.

It is the residual separation and zero waste research facility that represents the key interaction between community and industrial responsibility for a sustainable future. Through this facility the community can deliver the simple but important message to industry that "if we can't reuse it, recycle it, or compost it, industry shouldn't be making it. We need better industrial design and better consumer education for the twenty-first century."

In terms of communicating to the general public the need for better industrial design and the fact that we need to cobble together both community responsibility (at the back-end) and industrial responsibility (at the front-end)

FIGURE 2.12. Some of the items in the residual fraction in Capannori, Italy. The large number of one-use coffee capsules has triggered correspondence between Capannori's Zero Waste Research Center and the manufacturer. *Photograph by Pier Felice Ferri*

to get to zero waste, I strongly recommend that readers stress the fourth R of *redesign* to get this message across. We need to reduce, reuse, recycle, and redesign to get to zero waste.

An Institute for Zero Waste and Sustainability

The attraction of zero waste as a tool to advance toward sustainability is that every human being is involved with the problem every day. Every day we make waste we are part of a nonsustainable way of living on the planet, and every day we "unmake waste" by separating our discarded materials and by avoiding unnecessary products and packaging we are part of a sustainable way of living on the planet. But we have to institutionalize this move toward sustainability.

As these residual separation and zero waste research facilities begin to emerge they need to network with each other and with a larger and central entity (regional, national, or even international), which could be called the Institute of Zero Waste and Sustainability. Such an institute would not only aim to recommend better industrial designs to industry, but would also attempt to link zero waste with the other demands of a sustainable future. We need to integrate those

FIGURE 2.13. Zero waste goals can integrate with other developments needed to move toward a sustainable society. *Illustration by Jack Cook*

working on this issue with many other sectors in society. It is easy to see how this can be done: Composting can be linked to sustainable agriculture; anaerobic digestion can be linked to sustainable energy; deconstruction can be linked to green architecture; the residual screening and zero waste research facilities are clearly linked to education and better industrial design; the reuse and repair centers can be linked to community development; and the whole program can be linked to sustainable economic development and job creation (see figure 2.13).

So many sectors are involved in this holistic approach that it would be making a tragic mistake to leave all this to "waste experts." We need everyone involved in the greatest challenge of our age: moving toward a sustainable society. So instead of a few waste disposal experts we all have to become "resource management" experts, even if we only impinge on this system at the points of purchase and disposal.

Step 9: Better Industrial Design

There are three important developments industry needs to pursue: design for sustainability, clean production, and extended producer responsibility (EPR).

Design for Sustainability

Right from the outset industry needs to incorporate this new ethic. It is not enough for manufacturers to sell their products to the present; they must design them so that the object, or at least its constituent materials, can easily be shared with the future.[31] Packaging must be designed for reuse and products must be made for a prolonged life and capable of easy disassembly and repair. This is all part of what some have described as the triple bottom line or the three Ps: profit, people, and the planet.

Clean Production

Another important challenge in sustainable design is to eliminate as much as possible the use of toxic elements and compounds in manufacturing products.[32] This includes toxic metals like lead, cadmium, and mercury (which have no known biological use) as well as compounds containing the problematic elements chlorine, bromine, and fluorine. These halogenated compounds are so persistent that they come back to haunt us in many ways, either by reaching the stratosphere and damaging the ozone layer or by accumulating in our body fat, and being passed by pregnant woman to their unborn children and by breast-feeding mothers to their infants. Precursors for these substances (such as polyvinylchloride [PVC]) pose dangers to the workers involved in their manufacture, but their dangers are exacerbated when they are burned either accidentally in fires or deliberately in incinerators.

Extended Producer Responsibility (EPR)

Manufacturers and retailers can anticipate new laws that will force them to take back their products and packaging after the customer has finished with them.[33] This is already happening with laws requiring manufacturers to take back electronic equipment. The Waste Electrical and Electronic Equipment (WEEE) Directive is now European Union law.[34] Similar legislation is being proposed and enacted at the state level in the United States.[35] This legislation requires producers of electrical and electronic equipment to pay for the collection, treatment, and recovery of this equipment after the customer has finished with it. The EU law also requires Main Street shops and Internet retailers to allow consumers to return their waste equipment free of charge. Not only is this law keeping some very toxic materials out of the environment but it is also saving money when companies recover the gold and other valuable metals from these products.

An inspiring example of where EPR laws are working effectively is in Flanders, the Flemish-speaking region of Belgium (see chapter 10), where producers, importers, and retailers of certain items must by law not only take back waste products free of charge, but also meet collection and recovery targets. This obligation applies to a wide range of products—batteries, vehicles, and tires; printed matter; electronics; lubricating and industrial oil; lighting equipment; animal and vegetable fats and oil; and medicines.

Even before these kinds of "take it back" laws were passed, some manufacturers were well ahead of the game, having found that recovering and reusing either the parts or the materials in their products saves money on both disposal and production costs. When I visited the Xerox Corporation in Europe in 2000, I found that the company was picking up old machines and trucking them to huge warehouses in Venray, in the Netherlands. There they were separated into four groups: machines that could be used again with a little cleaning, machines that could be used again with a few key parts replaced, machines that could be disassembled for reusable parts, and machines that could be stripped down for recyclable materials. All the reusable parts were cleaned and tested to ensure that they met normal specifications. In the assembly plant, robots could be programmed to select either a used or new part as required. The final refurbished machines had to meet all the same quality controls as machines made entirely of new parts. At the time, the company was recovering 95 percent of the material in these products either as reusable machines, reusable parts, or recyclable materials. Even more impressive than the massive logistical feat involved to achieve this was the fact that, according to company officials, it was saving Xerox $76 million a year. This program is just part of the waste prevention and management programs of Xerox.[36] Gary Liss discusses some of the other companies that are also reporting huge savings when embarking on zero waste strategies in chapter 22.

Solid waste is the visible face of inefficiency. Those progressive companies that work on achieving zero waste save money both on production and disposal costs.

Step 10: Interim Landfills

Traditionally, the approach to resolving the problems posed by landfills has been to apply more and more sophisticated engineering to landfills in an effort to contain or capture both gaseous emissions and liquid effluents. This has involved daily cover, methane capture, and lining and leachate collection systems. Essentially, the goal has been to control what comes *out* of landfills, not what is put *into* them. However, as the US Environmental Protection Agency (EPA) has made clear, all landfills eventually leak.[37] So if we cannot control what comes out of a landfill, we must control what goes in. The zero

Dealing with Household Toxics

Household toxics such as paint, solvents, cleaners, batteries, and fluorescent bulbs only account for about 1 percent of the domestic stream—but it is a very important percentage since incorrect handling of it can come back to haunt us.

Communities have handled household toxics in a variety of different ways, both within the previous ten steps and in separate measures.

Household Toxics Collection Days. These tend to be very expensive to organize and often result in long lines of irritated customers.

Household Toxics Drop-off Sites. Halifax, Nova Scotia, has a regular Saturday morning drive-in drop-off site where citizens do not even have to get out of their cars.

Special Facilities. San Francisco operates a special facility for the handling of household toxics.

Ecodepots or Islands. Nova Scotia has about one hundred Enviro-Depots around the province that act as redemption centers for beverage containers with a deposit on them. Some of these also accept household toxics like batteries.

Door-to-Door Collection of Specific Items. In Neunkirchen, Austria, the municipality provides a bright red, screw-top plastic container designed to receive all household batteries. When full, these are set out with the recyclables and go to the recycling facility. With one simple movement

waste approach recognizes this truism and sets out to control what goes in to an interim landfill using the residual separation and zero waste research facility. I use the term *interim* because in the zero waste strategy the vision is to eliminate landfilling altogether. However, we still need to use an interim landfill for handling what we can't currently reduce, reuse, recycle, compost, or redesign out of the production system.

The interim landfill in the future will only receive a biologically stabilized dirty organic fraction and the current nonrecyclable fraction. If we can get very good at keeping both toxic and biodegradable materials out of landfills, we might be able to return to the notion of filling holes in the ground, like old quarries, without environmental complications. Meanwhile, with this screening approach we can certainly make landfills much smaller (see the Halifax landfill in figure 2.11) than raw waste landfills and a lot safer than incinerator ash landfills.

The task of those running a residual separation and zero waste research facility is to reduce the quantity, toxicity, and biodegradability of the material workers on the conveyor belt can pull out literally hundreds of small batteries very efficiently.

Stores Reusing Paints. In Halifax, Nova Scotia, and Canberra, Australia, stores have been set up to reuse paints.

Reuse and Repair Centers. These centers also frequently accept some toxic items like paints and oil for use in their reuse and repair operations.

Take-It-Back Systems. The city of Ottawa, Canada, has a system of retailers who take back various toxic products such as pharmaceuticals, paint, batteries, and computers. The city lists these retailers in booklets for the public. Many appreciate the fact that in addition to the positive public relations obtained, offering this service brings many prospective customers into their stores.

Pharmacies Accepting Old Medicines. Outside many pharmacies in Italy, a secure yellow container is available for the return of old medicines and containers.

Sometimes technical developments that look very promising for one reason can have unexpected negative consequences for other reasons. One such case is the fluorescent bulb. While these save energy and produce less waste heat than conventional tungsten-filament bulbs, they contain mercury. When these bulbs are broken, that mercury is released and can cause both health and environmental problems. So they must be handled with great care in waste disposal or recycling programs. Hopefully, the more recent energy-efficient LED lights will not have other unexpected toxic consequences.

we are forced to bury in this interim landfill. The goal of the Zero Waste 2020 strategy is that by 2020 there should be very little material we are forced to dispose of in this way.

Arguments for the Ten-Step Plan

For those skeptical readers who may be thinking that a lot of these suggestions sound like the idle fantasies of a retired professor, I urge you to jump to chapter 10 and examine all the initiatives that Flanders has put in place to solve its waste problem. Flanders is, in fact, undertaking almost all of the ten steps (except Steps 8 and 10).

There are several reasons why this ten-step plan should be attractive to citizens and decision makers concerned about the local environment and the local economy, as well as those who are concerned about the bigger picture. Ultimately, its acceptability is as much about psychology as it is about practicalities. Here are some of the advantages of the ten-step program to zero waste:

- It is low-tech: Most of the facilities can be designed and built by local companies and operated by local communities.
- Unlike incinerators, it keeps most of the money spent on the system in the local area, creating jobs and small business opportunities.
- Every component is operating somewhere in the world.
- It is better for health, better for the economy, and better for the planet.
- It introduces discipline into the system via residual separation and research.
- With the research center it integrates higher education into the struggle for zero waste and sustainability.
- It is positive; citizens will say yes to most of these rational steps.
- It brings people together instead of driving them apart (zero waste is above politics; it appeals to all parts of the political spectrum).
- It challenges the creativity of both citizens and decision makers.
- It offers more hope to our children about the future.

Zero Waste Is Being Co-opted

Unfortunately, in some places we are already seeing that the zero waste strategy is being co-opted and corrupted. Two words can completely undermine the effort to reach zero waste and a genuine move toward sustainability. Readers should beware of the phrase *zero waste to landfill*, which is a euphemistic way of accepting some form of incinerator into the mix.

However, megalandfills and incinerators are neither acceptable nor necessary. In this overview I have outlined the Zero Waste 2020 strategy, which is not only better for our health, but for the local economy and our planet.

There remains an obstacle, though—one that I call the bad law of pollution. When we compare communities, provinces, or countries, the level of pollution increases as the level of corruption increases. The more corrupt your community, the more polluted it will become. Fortunately, there is also the good law of pollution, which states that the level of pollution decreases as the level of public participation increases. In short, we need to clean up the political system in order to clean up our environment.

Nowhere is this corruption more apparent than the continued promotion of megalandfills and incinerators. A few people make a huge profit from building and running these entities, but the rest of us—not least our children and grandchildren—will foot the bill in the future in countless ways. In chapters 17 and 18, Buddy Boyd and Helen Spiegelman warn us about the too-cozy relationship between local government officials and the waste industry that favors dirty recycling, incinerators, and landfills over a genuine commitment to zero waste and sustainability.

However, the one good thing that comes out of these landfill and incineration battles is that it galvanizes the public into passionate opposition. It is this passion, and the networking it generates, which provides the push for the alternative zero waste strategy.

We need creativity to replace corruption. Today, nowhere is the struggle between the mindless political pressures for incineration and the countering citizen pressure for zero waste more apparent than in Italy. Several communities in the Naples area have already endorsed a zero waste strategy; the mayor of Naples announced in 2011 that Naples itself was going to embark on a zero waste strategy. Is this the kind of creative leadership that the world once saw from Italy during the Renaissance and the scientific revolution? Perhaps so.

A Brief History of the
Anti-Incineration Movement

To begin on a personal note, I was first introduced to the issue of incineration of municipal waste in 1985. I was only one and a half years into my teaching career at St. Lawrence University when the college's science librarian, Jane Eaton, asked me, "You're a chemist—what do you think about the proposed county incinerator?" I responded that I thought it sounded like a good idea because it would mean we could get rid of the thirty-two leaking landfills in our county and we could create some energy in the process.

Eaton said, "I think you had better read this," and handed me a paper on dioxin emissions from incineration, which had been written by Barry Commoner and his colleagues at the Center for the Biology of Natural Systems (CBNS) at Queens College in New York.[1, 2] It was a shock to read this paper because if anyone had asked me about dioxin before that moment I would have thought of Agent Orange, the notorious defoliant used in Vietnam, or industrial accidents like the one in Seveso, Italy. It was hard then (and still is) to believe that we can make these same dioxins—the most toxic substances we have ever made in a chemical laboratory—simply by burning household waste.[3]

One of the things I found most shocking about Commoner's paper was the revelation that one of the consulting engineers promoting incinerators (Floyd Hasselriis) had apparently manipulated the data to show in a graph that dioxin emissions smoothly fell as the temperature in the furnace rose.[4] When Commoner checked the original data measured in the Hamilton, Ontario, incinerator he found no such correlation existed.

In fact, I was shocked enough to phone Commoner. He told me that he was going to confront Hasselriis on this at an upcoming conference at Hofstra University and suggested that I come to witness the exchange. I did, with tape recorder in tow. At this meeting Hasselriis said that out of the thirteen original data points, he had used *four* data points to draw the line in his graph (which was drawn without any data points indicated). In his original paper he said that he had dropped two outliers, which meant that he had used *eleven* of the data points. A drop from eleven to four was bad enough, but when I checked the data later myself I found that only *two* of these four data points lay on the line in his report—the line so clearly implying that dioxin releases went down as incinerator temperatures rose. So, out of thirteen original data points Hasselriis had used just *two* to draw the line in his graph.

Essentially, by selecting two data points from thirteen available Hasselriis could draw any line he wanted. In this case the line he drew ran parallel to the carbon monoxide emission line. So it was very clear what Hasselriis was after. It is very expensive and complicated to measure dioxin emissions from an incinerator, but carbon monoxide levels can be monitored cheaply and continuously. If other scientists had accepted Hasselriis's fraudulent thesis it would have meant that the incinerator industry would not have to measure dioxin emissions—relying instead on their carbon monoxide emissions as a gauge. This would have relieved incinerator operators and regulatory bodies of the onerous and politically problematic task of measuring and reporting dioxin emissions. I wrote up my concerns about Hasselriis for a local environmental newsletter[5] and later presented the argument more formally in a paper presented at Clarion College in Pennsylvania.[6]

You don't have to lie about something that is good. Clearly, there was something very wrong with incineration dioxin emissions. When I brought this matter to the attention of the person in our local authority in charge of the environmental assessment for the local incinerator project, he said, "So one expert screwed up."

I replied, "He didn't just screw up. He lied."

He responded, "I have met recyclers who lie."

At this point I saw that we were in for a long battle. I realized from this exchange that I was no longer simply exchanging scientific information with this local official. It was clear that he had his marching orders. I inwardly groaned about what lay ahead, but little did I realize that my unexpected initiation into this issue was going to change both my life and my career in many ways.

Hasselriis was not alone among combustion engineers in the belief that the "dioxin problem" could simply be solved by good combustion and running incinerators at high temperatures. However, Commoner and his colleagues found that some of the highest dioxin emissions were coming from incinerators whose furnaces were operating at the *highest*, not the lowest, temperatures. Commoner hypothesized that dioxin was formed after combustion. This position was vindicated in 1985 when an Environment Canada team led by Vlado Ozvacic measured dioxin emissions from an incinerator on Prince Edward Island. They found no dioxins in the flue gas leaving the combustion chamber but did find them at the base of the stack, indicating their formation in the cooler parts of the system somewhere between the furnace and the stack.[7]

Ozvacic reported his findings at the Fifth International Symposium on Chlorinated Dioxins held in Bayreuth, Germany. I attended this conference and it was a real eye-opener as far as dioxin was concerned for me to witness firsthand the very cozy relationship among consultants, industry, and government. It was apparent that governments and industry (including the incineration industry) had a very strong interest in downplaying the significance of dioxin, but scientists who served these interests wanted to do so in a way that did not depress their research funding!

Not long after being exposed to the dioxin issue, my wife, Ellen, and I started a campaign to stop St. Lawrence County building a relatively small (250 tons per day) incinerator in Ogdensburg, New York, some seventeen miles from our home; we finally won this battle at the eleventh hour. After five and half years from the start of our opposition, the county legislature voted not to support a bond issue to finance the project.[8]

As we fought our battle in northern New York, we began to hear from other citizens fighting incinerators in nearby communities and states who sought our technical help. It was after a debate organized by the Vermont Law School in 1986, in which Barry Commoner and I took on two waste industry consultants (Kay Jones and Fred Lee), that—with activists from Connecticut, Massachusetts, New Hampshire, New York, and Vermont—we formed a coalition to fight incinerators nationwide. The first name for this coalition was explicit but impossibly long: the National Coalition Against Mass Burn Incineration and for Safe Alternatives. Eventually this cumbersome name morphed into Work on Waste USA, for which my wife edited the newsletter, *Waste Not*.

Before our involvement began in the early 1980s, when communities were confronted with incineration proposals they reached out to a number of organizations for help. This included the Institute for Local Self-Reliance (ILSR), which provided information for citizen groups, local decision makers, and local businesspeople on alternative strategies and technologies; the Citizens' Clearinghouse for Hazardous Waste, CCHW, led by Lois Gibbs of Love Canal fame (now Center for Health, Environment and Justice, CHEJ),[9] for advice on how to fight their battles politically; and the Environmental Defense Fund, whose scientists Ellen Silbergeld and Richard Denison gave important scientific advice on the toxicity of incinerator ash and some information on dioxin. Silbergeld coauthored a paper (with Alastair Hay) that indicated that Monsanto's doctors had fudged the data on their cancer study on the workers exposed to dioxin during an accident in Monsanto's plant in Nitro, West Virginia.[10, 11] However, much of the scientific information on dioxin used by citizens in incinerator battles was provided by Barry Commoner and his colleagues from CBNS, especially Tom Webster.

Establishing the Link Between Incinerators and Dioxin in Cows' Milk

Tom Webster and I went on to present papers on dioxin at six of the annual international dioxin symposia held between 1986 and 1997. These papers, subsequently peer-reviewed and published in the journal *Chemosphere*, focused on the buildup of dioxins and related compounds in the human food chain resulting from incinerator emissions.[12–17] This took on a special importance for us in St. Lawrence County because at the time we were the largest milk-producing county in the state. When the county hired consultants (Battelle out of Columbus, Ohio) to do a health risk assessment for dioxin emissions, they only looked at risks from inhalation and completely neglected exposure via the food chain. We subsequently found that cows' milk could expose an individual to nearly two hundred times more dioxin than inhalation. Indeed, by our calculation, one quart of milk would expose an individual to the equivalent of breathing the air next to the grazing cow for eight months.[18] About ten years later a study conducted in Germany, and based upon field measurements, confirmed that a grazing cow would put into its body in one day the same amount of dioxin that would be obtained by a human being breathing the same air as the grazing cow for fourteen years![19]

At the time, our theoretical calculations were dismissed by consultants, but later field measurements made of cows' milk downwind of an incinerator in Rotterdam validated our concerns. The government ordered that the milk of dairy cows in sixteen dairy farms downwind of this Rotterdam incinerator not be sold but instead be collected and sent by the government to a facility where the fat (where the dioxins are concentrated) was removed and then sent to a hazardous waste incinerator to be destroyed. The irony is that both the trash incinerator and the hazardous waste incinerator stood side by side and were operated by the same company. This must qualify as the weirdest form of recycling in history.[20, 21]

A few years later three incinerators operating near Lisle in France[22] had to be closed because of high dioxin levels in cows' milk in the area. Meanwhile, measurements made in cows' milk in Ireland, which had no trash incinerators at the time (but also a lot less industry), were about ten times lower than in the United Kingdom and about one-fifth of the "ideal" goal set by the German government.[23]

The Power Decade: Three Hundred Incinerators Defeated in North America

Between 1985 and 1995 grassroots groups defeated about three hundred of the nearly four hundred incinerators planned for the United States and Canada.[24] California had plans for thirty-five incinerators in 1985; they were only able to build three. New Jersey wanted twenty-two; they only got five. New York City was planning six incinerators—and they got none.

Work on Waste USA was instrumental in helping in many of these battles. We helped in five main ways.

Waste Not. We sent out weekly *hard* copies of *Waste Not*, as this was before the availability of the Internet. These helped keep citizens opposed to incineration informed of victories and defeats. This kind of news service was very important, because while it wasn't difficult to get local media coverage on incinerator battles, the national media never did a serious job of covering the issue. Too many journalists and editors (including those of the *New York Times*) swallowed the public relations spin of the incinerator industry that these were waste-to-energy facilities and represented the only viable alternative to megalandfills.

Those activists today who were not around in these ancient days—and whose activities are made possible today via the Internet—will find it difficult to contemplate all the work involved in getting out such a newsletter forty-eight times a year. First, it had to be composed and then typed out—and then corrections made with a fluid called Wite-Out. At this point a copying machine had to be found to run off several hundred copies. Then the copies had to be folded, stuffed into envelopes, and each envelope addressed and stamped and finally delivered to the post office. The whole exercise could take at least two days and was very expensive. Today, of course, one simply types a message and with one click on a computer it is delivered to hundreds of people all over the world. But either way, communities fighting incinerators then and now find communication is key to their efforts.

Videotapes. Beginning in 1986 as VideoActive Productions,[25] Roger Bailey (a professor of fine arts at St. Lawrence University) and I produced about forty videotapes on the dangers of incineration and on communities pursuing recycling, reuse, and composting. The most popular of these videotapes was the one titled *Waste Management: As if the Future Mattered*,[26] which was accompanied by a forty-eight-page booklet.[27] These videotapes also included a ten-part series on dioxin.[28]

Citizen Dioxin Conferences. The First Citizens Conference on Dioxin held in Chapel Hill, North Carolina, in 1991 was the first of a series of four Citizens Conferences on Dioxin in which we had great help and support from local organizers like the late Billie Elmore from North Carolina as well as Greenpeace and CCHW.[29] Not only were we able to provide a large number of grassroots activists a better understanding of the dioxin issue, but we were able to show honest scientists working for the EPA and other regulatory agencies that there was a large public interest in finding out the truth about dioxin and related compounds. This pressure was an important counterbalance to the pressures from various industries to downplay the significance of dioxin exposure to human health. The public relations stance of the chemical industry was that while dioxin was very bad for animals it presented little problem to humans. This view was best represented by an article by Fred Tschirley in *Scientific American*[30] to which I responded.[31]

Public Presentations. As more and more communities faced incinerator proposals, I traveled to forty-nine states to give presentations to communities. Since the early 1990s my work on waste has also taken me to sixty other countries. I have found it is critical to talk to community members, public officials, and sometimes to the industry itself; I did, for instance, at the Fourth International Management Conference on Waste to Energy in Amsterdam in 1998—right in the lion's den of the proincinerator waste industry (the International Solid Waste Association [ISWA]) itself.[32]

Debates. Along with others like Tom Webster and Barry Commoner, I challenged incinerator promoters and consultants to debate the issue in the communities where these facilities were being proposed. More often than not they refused, but when they agreed both citizens and decision makers were usually convinced that incineration was not the panacea that it was cracked up to be. By this method we were able to punch through the public relations hype of promoters and empower citizens to fight the battle themselves.

When I say that Work on Waste USA and other groups have helped communities defeat many incinerators with information and expert advice, it is important to stress that only communities themselves can actually achieve the victory; no one from outside can do it for them. As I have said in many public presentations, effecting change is like driving a nail through a piece of wood. The expert can sharpen the nail, but you need the hammer of public opinion to drive the nail home.

An open letter circulated by GAIA and a number of environmental justice groups sums up what it takes to beat incineration and other unwelcome projects pushed on communities by powerful corporations:

> To build a powerful movement, you must first figure out where you have power, and build from there. We have power in our communities where we have relationships and can hold politicians and corporations accountable. In DC, corporate power rules because they can concentrate energy and resources there—in ways we cannot. However, when confronting these same corporations in our tribes, cities, and towns, we reveal that they are not nimble or powerful enough to defeat our communities. Movements are built house-by-house, block-by-block, community-by-community, whenever people in communities rally around a common cause, acting on their own behalf with allies and networks—often against powerful interests, often building new institutions needed to win lasting change.[33]

Other groups and individuals that played a large role in fighting incineration in North America were:

Public Interest Research Groups (PIRGs). These groups were the brainchild of consumer activist Ralph Nader. The New York PIRG was particularly active in the fight against the incinerators proposed in New York City and other New York communities.

Local Chapters of the Sierra Club. These were very helpful in incinerator battles in Detroit, Michigan; Austin, Texas; Scranton, Pennsylvania; Portland, Oregon; and several other communities and states. The national group was rather slow in taking an official position against incineration but did so eventually.

Daniel Knapp and Mary Lou Van Deventer. Urban Ore was particularly effective at battling many incinerator proposals in California—especially in the San Francisco Bay area. They also played an important role in protecting the language used to describe recycling activities. Knapp and Deventer have always insisted that "it's not waste until it is wasted." They were early fighters against the waste industry's greenwashing. To this day incinerator proponents try to get incineration classified as a form of recycling.

Peter Montague. The whole antitoxics, anti-incineration, and antilandfill movement was also hugely helped by the weekly *Rachel's Hazardous*

Waste News distributed by Montague, director of the Environmental Research Foundation (ERF).[34] Indeed, it was this wonderfully succinct tool designed to help citizens on many technical issues that we used as a model for *Waste Not*. Each issue of *Rachel's* was short enough to be read on the day it was received. Everything was meticulously documented so that citizens could use the information in public hearings and meetings with confidence. Each issue had a large identifying number, the date, and was already punched with three holes for convenient filing and retrieval for later use. In fact, *Waste Not* copied this format and tried to do for news coverage on incineration and recycling battles what Montague did with technical advice on hazardous waste and many other issues. Both were designed to give citizens ammunition to fight the Goliaths of the municipal and hazardous waste industries. According to author Robert Gottlieb, "By the late 1980s, both the Connetts and Montague had become important adjuncts to the grassroots groups and anti-toxic networks.... Their publications became essential reading for community groups. They made obscure documents and reports accessible, covered project battles and revealed information the waste industry would rather have kept removed from public view."[35]

In the late 1980s and early 1990s these grassroots efforts eventually won the support of several national environmental organizations, including Greenpeace, the Sierra Club, the National Resource Defense Council (NRDC), and Clean Water Action, all of which joined forces in a campaign to fight incineration at the national level.

The Supreme Court Ruling on Flow Control

Another obstacle for incinerators came in the form of challenges to flow-control laws, which required private waste haulers to take waste to a designated facility. In *C&A Carbone, Inc. v. Clarkstown* (1994), the Supreme Court struck down the Clarkstown, New York, flow-control ordinance, stating that it violated the commerce clause in the US Constitution. Clarkstown had hired a private contractor to build a waste transfer facility and put a flow-control ordinance into effect that mandated all the town's waste had to be directed to the facility. But the court ruled that solid waste was a commodity and that restricting its origin or destination deprived competitors of access to a local market.[36]

The Supreme Court ruling had a huge impact in New Jersey, where garbage haulers were freed from taking their waste to expensive incinerators in the state at ninety dollars per ton and instead could truck their waste to landfills in Pennsylvania for half the tipping fee. This threw the five operating incinerators in New Jersey into financial disarray. They were forced to lower their tipping fees to forty-five dollars per ton to capture the waste they needed to operate and meet their contractual obligations to deliver electricity to their customers. However, with this lowered tipping fee they could no longer service their debt for the capital costs accrued building their facilities. They went to the state to bail them out, but the state refused.

Moreover, without the ability to capture the waste in this manner with "flow control" it made it very difficult for incinerator companies to get investors to put up the huge capital investment needed to build these plants in the future. A good example of the problem that this ruling caused the incinerator industry came in Mercer County, New Jersey. The county had planned a trash incinerator to burn over 1,600 tons of waste per day, but they canceled the project—which would have required twenty-year commitments for sources of trash in order to secure the necessary bonds. One of the reasons given for this cancelation was provided by the Delaware Riverkeeper, a local group opposed to the incinerator: "This is the worst time to invest bonds in a $260 million municipal trash system, especially if the sources of trash must be secured on a long-term basis. No other NJ county is taking this risk now."[37]

This Supreme Court ruling combined with the continuing unpopularity of the technology with the public resulted in no garbage incinerator being built in the United States between 1997 and 2010, although some older ones have been retrofitted and expanded. Others have been closed and some have incurred huge debts that have put cities into a precarious financial situation. The most well known example is Harrisburg, Pennsylvania, which filed for bankruptcy.[38] However, the Supreme Court reversed its position on flow control in 2007 in *United Haulers Association, Inc. v. Oneida-Herkimer Solid Waste Management Authority*.[39] This may explain the renewed interest in building incinerators again in the United States.

Behind the Glowing Reports from Europe

Unfortunately, in the battles against incinerators worldwide European governments have not been helpful. Historically, a number of European countries—particularly Austria, Belgium, France, Germany, the Netherlands, Denmark, Sweden, and Switzerland—have relied heavily on incineration. Because many—especially Germany, Switzerland, Denmark, and Sweden— have good reputations for strong regulations, reliable technology, and tough governmental enforcement, many decision makers in other countries have been lulled into a false sense of security as far as the safety and monitoring of incineration in their own countries is concerned.

In the 1980s and 1990s a number of communities in North America (about seventy) were persuaded to build incinerators based largely upon glowing accounts from political leaders who were sent on all-expenses-paid trips to European countries to visit their burn facilities. However, few of these politicians spoke to citizens and farmers who lived near these facilities, many of whom were far from happy with having these burners in their towns or operating near their farms. In fact, during the late 1980s and early 1990s there was a very strong grassroots movement against incineration in Europe, especially Germany.

In a videotape Roger Bailey and I shot in Germany and the Netherlands in 1990,[40] we interviewed farmers downwind of the large incinerator in Rotterdam and citizens in Bavaria who had organized a movement called Das Bessere Müllkonzept (the Better Waste Concept).[41] Das Bessere Müllkonzept was an alternative waste strategy, designed in response to a plan to build another seventeen trash incinerators in Bavaria. In retrospect, this plan looks very similar to the ten steps to zero waste outlined in chapter 2, except for the latter's focus on better industrial design and the need for residual separation and research facilities. The Bavarian plan would have eliminated the need for the new incinerators, and the citizens tried to get it passed as a "citizens' law." The first step was to get the issue on the ballot. This required 10 percent of Bavarian voters in a twelve-day period to sign a statement in favor of Das Bessere Müllkonzept at their town hall in front of a witness. This daunting task had never been achieved before on any issue. The grassroots organizers needed 850,000 signatures. Remarkably, with little help from the media, they were able to get *over one million* people to do this![42]

Unfortunately, when the vote was held, the citizens lost, but only by a very narrow margin. The citizens won all the major cities, but the government was able to win the rural areas. However, despite the loss it was a huge moral victory and very few of the seventeen new incinerators planned for Bavaria were ever built. It was also a very graphic example of the fact that Europeans were not in love with incineration, as they were so often depicted to be.

Today, European countries are achieving higher recycling rates than ever before and the end result is that several countries like Germany, Sweden, and the Netherlands are being forced to import waste from other countries to feed their incinerators.

According to Mathieu Berthoud, Northern Europe CEO for the giant waste corporation Sita, "Most German studies show an overcapacity [in Germany] estimated between 2 and 4 million tons.... Already today Germany is a net importer of waste in order to satisfy the demand of the incinerators." Citing declining waste quantities and the expansion of incineration capacity, some in Germany and the Netherlands, said Berthoud, are already calling for a moratorium on the extension of German incineration capacity. Sweden, where incineration capacity grew from 2 million to 6 million tons in the last ten years, faces similar issues. "In 2012 the overcapacity will reach 2 million tons," says Berthoud. "Today Sweden continues to build thermal facilities as waste incineration is seen as an energy production business. They are extremely dependent on imports. The situation can become worse if the imports from Norway are stopped."[43]

More and more political leaders are recognizing that even if you made incinerators safe you would never make them sensible because incineration wastes resources and does not take fundamental steps to reduce global warming. In other words, the argument today against incineration is less about toxicity (although there are some unresolved issues on that front) and more about resource conservation and sustainability.

Still other news emerging from Europe shows that the region's push for incinerators may be ending. A recent vote in the European Parliament supports the notion that no material be burned in incincrators that can be recycled or composted. Since this amounts to 90 percent of the discard stream this would virtually rule out the building of new incinerators in Europe and probably force the closure of many existing plants.[44]

Trends in Asia

At one point in the late 1990s Japan was operating three times more trash incinerators (about 1,800) than the rest of the world put together (about 600). Tokyo alone has twenty-three municipal waste incinerators. By local law in Tokyo each borough is required to be self-sufficient in terms of waste management, including the financial district, which means with so little space each has been forced to build an incinerator.

Not only did this obsession with incineration give Japan the dubious distinction of being the largest emitter of dioxins into the air in the world in 1995,[45] but to make matters worse Japan has aggressively marketed its incinerator technology throughout Asia. It's a three-step process. First, Japanese development agencies provide the waste plans, with an incineration plant being a central feature. Second, Japanese banks then provide the loans; third, Japanese companies build the plants. Targeted countries have included Indonesia, Malaysia, the Philippines, South Korea, Taiwan, Thailand, and Vietnam. Fortunately, strong citizens' movements have sprung up in most of these countries to resist many of these projects. One of the strongest resistance movements was in the Philippines, which in 1999 became the first country in the world to ban incineration.[46] In 2000, with the help of Greenpeace, the various grassroots groups in these countries came together to form an umbrella network named Waste Not Asia.[47]

China, on the other hand, seems determined to repeat the Global North's bad example of waste incineration. Beijing has nearly completed the building of Asia's largest trash incinerator. It will burn three thousand tons a day, which is about one-quarter of Beijing's waste.[48] According to a recent *Guardian* article:

> China now generates over a quarter of the world's garbage, at least 250 million tonnes annually. With municipal solid waste (MSW) growing 8% to 10% annually, cities are under great pressure to deliver advanced waste-management solutions. . . . Presently, incineration is growing at a feverish pace.[49]

According to another media report, China is currently building 150 to 300 incinerators, but some of them are receiving strong opposition from citizens' groups.[50]

India, too, seems to be following the incineration path. What is very disturbing about these new incinerators in China and India is that the capital costs are about ten times less than European or North American plants. It is hard to believe that such a difference is entirely due to lower labor and material costs; it more likely reflects the use of less expensive air pollution control equipment. If so, both countries might have to wait twenty years to discover the real costs of cutting such corners when they start to measure the levels of dioxins in their food and toxic metals in their soils.

The Worldwide Effort to Fight Incineration

In 2000 in South Africa I joined many grassroots activists from around the world who were fighting incinerators in their home countries for the formation of GAIA. This organization set up its headquarters in the Philippines and also maintains an office in Berkeley, California. On its website, GAIA describes itself as "a worldwide alliance of more than 650 grassroots groups, non-governmental organizations, and individuals in over 90 countries whose ultimate vision is a just, toxic-free world without incineration."

One of GAIA's key functions is to help network, via the Internet and the occasional international conference, the many grassroots groups fighting incinerators around the world so that they can help each other by sharing experiences, knowledge, and resources. Sometimes just simply knowing you are not alone lends a great deal of strength to a group's massive uphill battle against this technology. Incinerator proponents include government agencies, consultants, and well-heeled lobbyists.

In addition, GAIA has published invaluable reports on incineration, recycling, the importance of ragpickers in developing economies, and the role of the zero waste strategy in fighting global warming. All of these reports can be downloaded online without charge, and GAIA's website (www.no-burn.org) is a goldmine of information for anyone fighting an incinerator proposal today and/or wanting information on the alternatives.

Back in the United States: Incinerators and Greenwashing

As Mark Twain may have said, "History may not repeat itself, but it does ·rhyme!" Today, we are again seeing efforts to build incinerators in North America. However, because their capital and operating costs are so high they cannot compete on a level playing field with recycling and composting, let alone a complete zero waste strategy. Thus promoters are currently seeking government subsidies by claiming that incineration is a "green energy" alternative to fossil fuels.

These claims are receiving support from a group of engineers at Columbia University who have strong ties with, and funding from, the incinerator industry. One of these engineers is Nickolas Themelis, who has given testimony in several states in support of subsidies for incineration in the name of "renewable energy." Here is part of his testimony before the Connecticut legislature in support of an act that would reclassify energy-producing incinerators as Class I renewable energy sources, and thus open the door for renewable energy subsidies:

> I am director of the Earth Engineering Center of Columbia University. . . . Our studies have shown conclusively that after all possible recycling and composting are done, the only two alternatives for dealing with the post-recycling municipal solid wastes (MSW) are combustion with energy recovery (also called waste-to-energy) or landfilling. Therefore, waste-to-energy (WTE; or "Trash-to-Energy") is the only source of renewable energy that also avoids the environmental impacts and land use of landfilling. . . . The proposed act is eminently fair in recognizing that WTE is also a renewable energy source and should be encouraged as much as solar and wind

energy. In March 2011, I was privileged to testify on behalf of similar legislation in Maryland that by now has become the law of that state. WTE power plants avoid landfilling, are sources of renewable energy and reduce the greenhouse gas (GHG) emissions of the state.[51]

On its website Columbia University's Earth Engineering Center (EEC) lists among its sponsors several incinerator companies, such as Covanta Energy, Energy Answers International, Martin GmbH, and Wheelabrator Technologies. Also listed are two consulting companies known for pushing incinerators—Gershman, Brickner, and Bratton, Inc. and HDR—as well as the proincinerator trade association Energy Recovery Council. It also includes the Plastics Division of the American Chemistry Council, the Solid Waste Association of North America (SWANA), and the EPA's Office of Solid Waste.[52] All of these organizations have a long history of supporting incineration. In addition, the research associates listed by EEC contain a veritable Who's Who of proincineration experts and consultants.[53]

Themelis is also the cofounder, with Maria Zannes (of the International Solid Waste Association, ISWA), of the Waste to Energy Research and Technology (WTERT) Council. Its mission statement is as follows: "The Waste-to-Energy Research and Technology Council (WTERT) brings together engineers, scientists and managers from universities and industry. The mission of WTERT is to identify and advance the best available waste-to-energy (WTE) technologies for the recovery of energy or fuels from municipal solid wastes and other industrial, agricultural, and forestry residues."[54]

WTERT's sponsors look identical to the sponsors of the EEC.[55] On its home page WTERT treats us to a series of revolving pictures of incinerators from around the world starting with Brescia and then Vienna. Below these pictures is a whole series of news items slanted toward incineration promotion. On March 1, 2012, Themelis even posted a short piece downplaying the dangers posed by dioxin. "As a result of the new controls," he wrote, "the dioxin/furan emissions of WTE plants in the United States and worldwide were decreased by over one thousand times to insignificant levels: Right now, all US WTE plants (28 million tons of MSW combusted) emit less than six grams TEQ* dioxins per year. In comparison, "backyard barrel burning" is estimated by EPA to be about 500 grams TEQ."[56]

---

* TEQ, or toxic equivalent, is a scale in which all the dioxin and furan family members have their toxicity compared to the toxicity of 2,3,7,8 tetrachlorinated dioxin, the most toxic of all the dioxins studied.

The reader should consult chapter 4 to see the flimsy basis on which such "optimistic" conclusions are based.

Also pushing for reclassifying WTE plants as renewable energy sources is the Energy Recovery Council (ERC), which identifies itself on its website as follows:

> The Energy Recovery Council is a national trade organization representing the waste-to-energy industry and communities that own waste-to-energy facilities. Current ERC members own and operate 69 of the 86 modern waste-to-energy facilities that operate nationwide, safely disposing of municipal solid waste, while at the same time generating renewable electricity using modern combustion technology equipped with state-of-the-art emission control systems. These facilities have been recognized by EPA as a "clean, reliable, renewable source of energy" that produce "electricity with less environmental impact than almost any other source of electricity." Energy Recovery Council members include Covanta Energy Corporation, Wheelabrator Technologies Inc., and Babcock & Wilcox, as well as 28 municipalities that are served by waste-to-energy plants and other associate members that work in the municipal waste management and energy fields.[57]

When most citizens see a statement from a trade association like this, I think they probably expect a very "rose-spectacled" view of the technology. And this is precisely what we get on the ERC's website when they tell us that "waste-to-energy produces clean, renewable energy" and elaborate:

> Waste-to-energy meets the two basic criteria for establishing what a renewable energy resource is—its fuel source (trash) is sustainable and indigenous. Waste-to-energy facilities recover valuable energy from trash after efforts to "reduce, reuse, and recycle" have been implemented by households and local governments.[58]

Contrary to these claims, I hope that readers were convinced by what was written in chapter 1—in no way can trash be regarded as a "renewable" resource, nor can incineration be seen as a sustainable technology (when you burn something, you have to go all the way back to square one and extract more virgin resources again to replace it). However, if you work for the incineration industry or promote it, what else are you going to say? One doesn't usually look to trade associations for objective assessments about their products. However, I think we should expect a higher standard of objectivity when it comes to the academic community.

When the average citizen hears the testimony of Themelis and others from the EEC or the WTERT Council, they will probably believe that they are hearing from independent academic sources. However, they are actually hearing from those with very strong ties to the incinerator industry and others who promote its interests. I think it is regrettable that the authorities at Columbia University have allowed their institution's name to be used in this way. But the activities of Themelis and colleagues clearly indicate that the incinerator industry is gearing itself up for a major push to build more incinerators in the United States after a hiatus of over fifteen years.

The industry's strategy will clearly be to emphasize energy recovery and search for subsidies from state and federal governments in the name of "renewable energy." Politicians desperate to find any way to get off Arab oil might be too easily convinced that this technology represents a win-win solution (get rid of your garbage and save energy as well). This seductive message fooled me when I first heard it in 1985 and doubtless will fool many others in the future.

Whatever local pollution and global warming issues are raised by citizens will be countered with the limited comparison to megalandfills. We have seen industry's trap of using the false paradigm of "either you bury it or you burn it" operate for over thirty years. Promoters will also claim that they support recycling and composting, and they are only interested in burning the residuals. In the 1990s this approach was called "integrated waste management." My response at the time was that if you believed the industry's line on integrated waste management you would believe that if you put a pig and a tiger in bed together you would get tiglets! What you get are little bits of pig. In other words the politicians in the pocket of the incinerator industry will give us lip service in support of pretty recycling programs while they sink the majority of the budget into hugely expensive incinerator projects. Instead of working to reduce waste, officials will be trying to find ways to, so to speak, feed the beast.

Sadly, some decision makers are falling for this new hype for incinerators. For example, there have been strenuous efforts over the last few years to build an incinerator in Frederick, Maryland, but the efforts are meeting very strong resistance from local citizens and environmental organizations.[59, 60] And as I write, the first mass-burn incinerator to be permitted in the United States since 1995 is being built in Palm Beach, Florida.[61]

The View from Canada

Several other incinerators are being proposed in Canada, and one is being built in the Durham region near Toronto—representing a huge setback for citizens in Ontario who have been supporting a zero waste strategy.[62, 63] Indeed, there was loud fanfare when it appeared that Toronto had declared such a policy in 2008.[64]

The successful effort to site an incinerator in the Durham region illustrates the importance for the incinerator industry of finding a local political champion with a lot of clout.[65] The American company Covanta Energy found one in the form of former police officer Roger Anderson, who is the powerful chair of the Durham regional council. Anderson has spearheaded the building of this massive incinerator to serve both the York and Durham regions.[66] Despite prolonged and determined resistance from citizens as well as from the local chapter of the auto workers union (CAW), the York-Durham regions voted in Anderson's pet project.[67] The auto workers union organized very large meetings in which I and others spoke about the viable and sustainable alternatives. But all to no avail—with Roger Anderson's leadership the incinerator project steamed ahead.

What has angered citizens most about Anderson's enormous influence in this matter is the fact that he does not even hold an elected position and thus is immune from the voters' wrath. He seems to have his colleagues under a hypnotic spell as the region has sleepwalked toward this facility and ignored the news that San Francisco has achieved 80 percent diversion from landfill without using incineration—again, more than incineration's 75 percent diversion rate (as the process leaves behind 25 percent in ash).

Even though enormous amounts of taxpayers' money will be spent building and operating this facility, when ground was broken for the site in 2011 with a lavish ceremony (again funded by taxpayers) local citizens were excluded.[68] This ceremony was voted number twenty-one in a list of "99 stupid things the government spent your money on" in *Maclean's*.[69] In my view citizens are going to find out in the future that it wasn't just the ceremony that was stupid but the building of this costly and nonsustainable project itself.

One of the arguments used for pushing this facility on the Durham region was the proclaimed need to be self-sufficient in waste disposal. Ironically, however, the ash from this facility (at least one ton for every four tons of waste burned) will be sent to a landfill in New York State.[70]

Despite proclaiming a zero waste strategy, Vancouver is also proposing one or more incinerators for the area (see chapter 9). In addition to these proposals for conventional incinerator projects in Canada, there have been a flood of smaller proposals for "incinerators in disguise." A number of names are used to describe these technologies, including gasification, pyrolysis, and plasma arc facilities. In each case the technologies use two steps. In the first step the solid waste is converted to a gas using a high temperature heat source. In the second step the gas (with or without cleaning) is then burned to produce energy. It is the second step that leads opponents to call these facilities "incinerators in disguise"—more appropriate names would be gasifying incinerators, pyrolyzing incinerators, or plasma arc incinerators. The group Greenaction in California has been very effective at exposing the sales hype and poor or nonexistent track records of the companies offering to build these "magic machines"—and along with GAIA issued a report on them.[71]

A classic case of sales hype winning the day over diligent research comes from Ottawa, the capital of Canada. Ottawa looked all set to move in a zero waste direction until the mayor and council got involved in the proposal to build an experimental plasma arc incinerator in the city by a company called Plasco Energy Group. Plasco is owned by Rod Bryden, the former owner of the city's ice hockey team, the Ottawa Senators. He had a close relationship with the mayor at the time his project was moved forward. Executives in the company also appear to have some other financial links to decision makers.[72]

Incredibly in December 2011, despite Plasco's dismal track record (in over three years of trials it has generated very little energy), the Ottawa city council voted 22–1 to sign a long-term contract to supply waste to the facility.[73] Worse, the council is going to do more than just host the facility—it also appears willing for the city to become a partner in the project, a fact that Plasco uses extensively in its promotion of proposals to build similar facilities elsewhere in North America. It is very disappointing to see Ottawa, which in so many ways is a progressive city, sully its reputation in this way.

When Los Angeles was considering using the same technology, the city's waste department sent a team to visit the Plasco facility in Ottawa. The operators tried to start up the facility (which had only been operating intermittently) *three* times, but failed to do so. The Los Angeles team left in disgust and cancelled any further interest in the project.

And Now New York City?

When I first got involved in the incineration fight in 1985, the leading battle was the attempt to build six incinerators in New York City. The first one was earmarked for Brooklyn Navy Yard. This produced intense resistance, which went on for many years. Leading the technical fight was Barry Commoner and CBNS and leading the political battle was the New York Public Interest Research Group (NYPIRG). Eventually, none of the incinerators were built. But once again history is beginning to rhyme. Recently, incinerator proposals have resurfaced in the city.[74] Hopefully, the council will be persuaded to spend a little money on air tickets for the mayor and his advisers to go to San Francisco and see how it has reached 80 percent diversion from landfills without using any incineration at all. This would be a far better investment than the millions of taxpayers' dollars that will be spent promoting this wasteful and unsustainable technology. This money will not solve New York City's waste problem, but it will make a number of consulting companies and law firms a lot richer.

The Trojan Horse

So sadly, once again citizens in North America in targeted towns are being forced to spend a huge amount of their time trying to persuade legislators that incineration is a Trojan horse as far as both the local economy and the global environment are concerned. For those, like myself, who were involved in these battles twenty-eight years ago it is horribly frustrating to have to win these arguments all over again with new elected leaders when we would rather be working with them to achieve the more rational and sustainable alternative of zero waste.

Fortunately, this time round we have the worldwide experience of GAIA to draw upon in fighting the new incinerator proposals. As GAIA spokesperson Ananda Tan puts it, "I have come to visualize our landscape as being like a giant 'whack-a-mole' game—where grassroots, community efforts succeed at beating down new burn proposals, only to have them pop up in a different community at a later point in time. The fact that no new (commercial-scale) incinerator has been built (in the United States and Canada) in over fourteen years is a testament to all the amazing local organizing and advocacy efforts." But the fight isn't quite like it used to be. Says Tan, "The industry is coming back with a vengeance, and this time they are more organized and better financed than before. This time around, they are creating political conditions that serve to undermine our local efforts quite severely, creating political and financial barriers for local zero waste initiatives."[75]

Here are the key elements Tan sees in the incinerator industry's expansion strategy:

Creating regulatory loopholes that further weaken federal and state mandates for protecting public health and the environment. Examples of this include efforts to redefine solid waste as a renewable resource and to reinterpret the way incinerator emissions are calculated under biogenic carbon (as opposed to synthetic organic materials made from fossil fuels like plastics) accounting standards, to make waste-to-energy (WTE) plants appear carbon neutral.

Securing more subsidies and policy incentives for both old and new incinerators.

Creating favorable investment conditions by convincing venture capital, energy investors, and banks that WTE is a significant and sensible renewable energy option.

Greenwashing their image through the sponsorship of environmental projects, support for corporate recycling initiatives, the creation of environmental justice (EJ) engagement tools, deploying allies in academia and federal agencies.[76]

As Tan rightly notes, grassroots community groups will be called upon to take on a growing number of community fights, but without the resources and capacity to take on the industry on all its fronts. Collaboration and networking among the hundreds of community activist groups will be key to mobilizing the kind of efforts necessary to win individual battles and keep a larger goal in mind: an end to incineration and move toward zero waste goals.

What Box?

When I first got involved in the incinerator battle, I wondered into what box incinerator promoters would put me. I had seen how they had handled Barry Commoner—renowned cellular biologist, writer, and environmental activist. They claimed he was not a real scientist because he had run for president. He was a politician instead. Other experts were dismissed because they were being paid for their testimony. With this tactic once you have labeled your opponent—in other words, put them in a box—you can throw away everything they say with just one line (the label on the box). Convenient. However, I was not running for political office and I was not being paid, so what box was I going to be put in? Eventually my label became clear: I was being called an "environmental evangelist." I felt at the time that I could live with that; the environment could do with some evangelism.

I was less happy when I heard I was being described as the "Jimmy Swaggart of trash." One day I dreamed up an answer to this: "God recycles, the Devil burns." This particularly incensed a proincinerator engineer who worked for the New York State Department of Environmental Conservation and who I believe was fairly devout, and I don't think he ever forgave me. However, I have used it in religious settings elsewhere without causing offense. In Acerra, near Naples, the bishop particularly enjoyed the joke and reminds me of it each time I meet him.

One particularly memorable occasion where I used this epithet occurred in Almadén, Spain, where I had a debate with the chief engineer of the mercury mine there (which has been operating for over two thousand years) who wanted to build an incinerator to burn imported hazardous waste. At the end of my talk a particularly irate member of the audience advanced toward me waving his finger and shouting that I "sounded more like a priest than a scientist!" Sensing the audience was on my side I rose to my feet and gave the sign of the cross and said, "That's right, my son—God recycles, the Devil burns." The crowd erupted—but positively, I am pleased to say. It turned out that my accuser was the company doctor who had been telling the workers for years that mercury was harmless for their health!

The next time I visited Spain I was met by a Greenpeace worker who proudly wore a large button with the legend *Dios recicla, el diablo quemadura.*

Incineration:
The Biggest Obstacle to Zero Waste

It is hard to understand why any rational official living in the twenty-first century and facing the critical need to develop sustainable solutions would countenance the squandering of finite material and huge financial resources on a nonsustainable practice like incineration. One European report has estimated that a combination of recycling and composting lowers greenhouse gas production forty-six times more than an incinerator producing electricity.[1]

Incineration might make sense if we had another planet to go to, but without that sci-fi escape it must be resisted in favor of more down-to-earth solutions that we can live with—both within our local communities and on the planet as a whole. Both incineration and landfilling attempt to bury the evidence of an unacceptable throwaway lifestyle. Every incinerator built delays this fundamental realization by at least twenty-five years—about the time it takes to pay back the huge capital costs involved in building the facility, *and* during that time it has to be fed, leaving little room to allow for more sustainable solutions to coexist.

Argument 1: Incinerators Are Very Expensive

Incinerators remain formidably expensive, but that expense is often hidden from public view with giant public subsidies. To pay for the capital and oper-

Ten Arguments Against Incineration at a Glance

1. Incinerators are very expensive.
2. Incinerators create very few jobs.
3. Incinerators are a waste of energy.
4. Incinerators are inflexible.
5. Incinerators produce a toxic ash and do not get rid of landfills.
6. Incinerators produce very toxic air emissions.
7. Incinerators release very toxic nanoparticles.
8. Incinerators are extremely unpopular with the public.
9. Incinerators are not sustainable.
10. There are better alternatives.

ating costs, as well as the operators' profit margins, the community or region will have to sign put-or-pay agreements, which trap them for twenty-five years or more. As the industry has struggled to make incineration safe, it has priced itself out of the market—or it would have if the market was applied on a level playing field.

Over half the capital cost of an incinerator built today goes into air pollution control equipment. Ironically, if the waste were not burned in the first place this hugely expensive equipment would not be necessary, nor would the toxic ash collected in these devices have to be sent to an expensive hazardous waste landfill, nor would the air emissions be subjected to very costly monitoring. But the public is being kept ill informed about the poor economics of incineration. Instead, they are being told that incineration is going to save their communities money.

For more information on the cost comparisons between incineration and intensive recycling and composting programs, see chapter 21.

Argument 2: Incinerators Create Very Few Jobs

Despite the massive investment involved, incinerators create very few long-term jobs for the community. Most of the money spent on incinerators goes into purchasing complicated equipment and leaves the host community and often the host country as well. It's extraordinary that during the current economic crisis in Europe, with the massive cutbacks in many public service areas and the concurrent loss of jobs, countries like the United Kingdom are still forging ahead with extremely costly, job-poor incinerator projects. For example, an incinerator approved for Blakenham in the county of Suffolk in the United Kingdom with lifetime costs of over $1.5 billion will create only forty-three permanent jobs.[2–4] However, there are many short-term jobs created in building these facilities and that is why they often attract local building union support. Not all unions are so shortsighted. In the United States the Teamsters union supports zero waste because their enlightened leadership sees a far greater job-creating potential in the alternatives to both landfills and incinerators.

One of Italy's most famous incinerators was built in Brescia (see figure 4.1).[5, 6] It cost about $400 million to build and has received at least another $600 million in energy subsidies. To my surprise, during a site visit to this incinerator, I was told that despite all the taxpayer money spent on this facility it had produced only eighty jobs. This is both economic and social madness—$1 billion to produce just eighty jobs!

FIGURE 4.1. The billion-dollar incinerator built in Brescia, Italy. This has been described by some incineration proponents as "the most beautiful incinerator in the world." *Photograph by Pier Felice Ferri*

Compare that with Nova Scotia (population 900,000), which avoided building a massive incinerator in Halifax and has created one thousand jobs in the collection and treatment of the discarded materials in the province and another two thousand jobs in the industries that use these secondary materials.[7] Or consider the fact that Recology—San Francisco's primary recycling, composting, and waste company—employs one thousand workers who are both unionized and employee owners of their company. As Robin Murray, a professor at the London School of Economics, points out in his book *Creating Wealth from Waste*, a zero waste strategy has extraordinary benefits for local economies.[8] No matter how hard the industry tries to claim otherwise, the same cannot be said for incinerators.

Argument 3: Incinerators Are a Waste of Energy

The argument that burning waste can be used to recover energy makes for good sales promotion, but the reality is that if saving energy is the goal, then more energy can be saved by society as a whole by reusing objects and recycling materials than can be recovered by burning them. Unfortunately, this argument is often lost on local decision makers who focus on energy gained locally and ignore the net loss nationally or globally. A combination of recycling and composting saves three to four times more energy than generated by an incinerator producing electricity.[9, 10] Some of the comparisons for individual materials are staggering. For example, recycling polyethylene terephthalate (PET) plastic (commonly used in disposable water bottles) saves twenty-six times more energy than burning it (see table 4.1).[11]

Table 4.1. Energy Comparison: Recycling versus Incineration (ICF Consulting, 2005)

MATERIAL	ENERGY SAVINGS FROM RECYCLING GJ/TONNE	ENERGY OUTPUT FROM INCINERATION GJ/TONNE	ENERGY SAVINGS (RECYCLING VERSUS INCINERATION)
Newsprint	6.33	2.62	2.4
Fine paper	15.87	2.23	7.1
Cardboard	8.56	2.31	3.7
Other paper	9.49	2.25	4.2
HDPE	64.27	6.30	10.2
PET	85.16	3.22	26.4
Other plastic	52.09	4.76	10.9

TABLE 4.1. A comparison of the energy saved by recycling with the energy released by burning and conversion to electricity. A greater net energy production occurs when waste heat is captured in combined heat and power incinerators.

Argument 4: Incinerators Are Inflexible

As a result of the huge expense incurred when communities build incinerators it greatly reduces their options for the future, no matter what developments and changes are happening elsewhere in the waste field. Such developments include the changing attitudes of the public toward recycling and sustainability and the changing value of the recovered materials themselves (see chapter 21). Their market value is expected to greatly increase with the expanding economies of India, China, Brazil, and other countries with rapidly developing economies and huge populations.

As Ludwig Kraemer, former waste director for the European Union, said in an interview on the BBC news program *Panorama* in 2000, "An incinerator needs to be fed for about twenty to thirty years, and in order to be economic it needs an enormous input from quite a region. So for twenty to thirty years you stifle innovation, you stifle alternatives, just in order to feed that monster which you build."[12]

In 1998, when the UK's Kent County entered into a twenty-five-year contract to burn 320,000 tons of waste per year, it thought it was making a wise economic move. But now, as the recycling economy has vastly improved, the county is losing an estimated $1.5 million a year. Rather than selling its recyclables for reuse—which would be both economically and environmentally

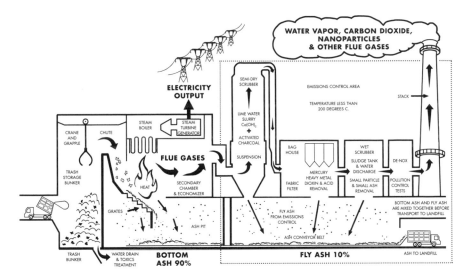

FIGURE 4.2. A schematic diagram of a modern incinerator showing where the bottom ash and fly ash are produced and collected. For every four tons of waste burned at least one ton of ash is produced. *Illustration by Jack Cook*

efficient—it must send those valuable resources to the incinerator just to meet its annual quota, an unfortunate situation that will persist until the contract expires. Said Keith Ferrin, the county council's environment spokesman, "What seemed a very wise decision a very long time ago is a very stupid one today."[13]

Argument 5: Incinerators Produce a Toxic Ash and Do Not Get Rid of Landfills

Incinerators don't solve the landfill problem. About 25 percent by weight of the incoming waste is left as ash, and this still has to be landfilled. Promoters of incineration often describe the ash produced as "inert." What they mean by this is that it is biologically stable (most of the organics have been burned away), but what is often implied is that it is nontoxic, which is false. The end result is that incineration converts four tons of trash into at least one ton of toxic ash that nobody wants.

Ash is a catch-22 for the incineration industry. As the industry has sought better air pollution control devices to capture the extremely toxic by-products of combustion, the resulting residues have become more problematic and costly to handle, dispose of, and contain.

There are two kinds of ash generated by an incinerator: the bottom ash (about 90 percent of the ash), which falls through the grate system at the base of the furnace, and the fly ash—the very fine material entrained in the flue gas. Ideally, this fly ash is captured in the boilers, the heat exchangers, and the air pollution control devices (see figure 4.2); however, inevitably a small fraction escapes into the atmosphere. As far as toxic metals are concerned, it is a chemical truism to state that the better the air pollution control the more toxic the fly ash becomes. The bottom ash is also toxic.

How this ash is handled varies dramatically around the world. In some jurisdictions like Ontario, Germany, and Switzerland, the fly ash is assumed to be a highly toxic material and *without testing* is automatically sent to hazardous waste landfills or containment facilities such as salt mines. In Japan, many of the incinerators vitrify the fly ash, and some even vitrify the bottom ash. However, the vitrification process uses up a great deal of the energy produced by the facility and is not popular with incinerator builders elsewhere.

In the Netherlands the bottom ash is used in road building and over 30 percent of the fly ash is used in asphalt. In short, when you drive a car in the Netherlands, you are driving over their ash landfills! In Denmark all the ash produced in their many incinerators is sent to Norway. In Norway it is dumped in a landfill on an island off the coast near Oslo.

In the United Kingdom the fly ash goes to open landfills with little control of the dust that flies off the surface. In one notorious ash landfill located near Gloucester and featured in the movie *Trashed*,[14] government bureaucrats tried to persuade citizens who lived nearby that the fly ash did not go beyond the fence line (see figure 4.3). Citizens were able to show this was nonsense when they produced pictures of a road sign on a nearby road, which had a thick layer of dust completely covering the message on the sign (see figure 4.4).

In some places fly ash has been used to make concrete, but with no warning on the product label that it contains hazardous materials like toxic metals and dioxins. One company—American Ash Recycling Corporation—that used incinerator ash in building products went bankrupt. Customers were rightly suspicious of using this toxic-containing material in buildings or other applications.

In the United States the toxicity of fly ash is being obscured by three things:

- the mixing of the fly ash with the bottom ash before testing;
- not testing for the absolute levels of toxics like metals and dioxins in the ash, but rather only looking at what dissolves out of the ash during a leachate test involving an aqueous medium; and
- the interference of the lime present in the ash with the current leaching test procedures.

FIGURE 4.3. Residents living around this fly ash landfill, near Gloucester in the United Kingdom, complained that fly ash was escaping the facility. *Photograph by Caroline Marsh*

FIGURE 4.4. A road sign covered by dust (presumed to be fly ash) from a nearby ash landfill site near Gloucester in the United Kingdom. Before this image was presented as evidence, government spokespersons had claimed that the airborne ash did not cross the facility boundaries. *Photograph by Steve Stone*

In fact, the current US ash testing regulations are essentially a giveaway to the incinerator industry. Starting in the 1980s the US incineration industry did everything it could to avoid any classification of ash as hazardous waste, which would have required either the fly or bottom ash (or both) to go to hazardous waste landfills. In 1986, David Sussman, the vice president of Ogden Martin, a major incinerator company in the United States at that time, summed up the situation this way: "It means finito, morte, the end for the resource recovery industry if ash is treated as hazardous waste. . . . Either that or widespread violations. There is simply no room for four million additional tons annually of ash waste. It would overwhelm all existing hazardous waste fills."[15]

For about ten years the incinerator industry, and its supporters, used a battery of arguments in an effort to avoid either the fly or the bottom ash receiving a hazardous waste classification. These included legal, linguistic, and chemical arguments.

The linguistic argument was used in New York State, where the Department of Environmental Conservation (a close ally of the incinerator industry in the 1980s and 1990s) declared the ash "special"—not "hazardous"— waste even when 50 percent of ash produced in New York incinerators was found to be failing hazardous waste tests.[16] Barry Commoner has called such tortuous argumentation "linguistic detoxification."[17]

The chemical argument maintained that it was not necessary to test the ash with acid in a leachate test (the test at that time was called the extraction procedure toxicity test, or EP tox test) because the ash was destined to go to a section of the landfill with no contact with regular waste, which generates *acid* leachate. However, when ash from the Claremont, New Hampshire, incinerator was tested with water instead of the acid medium in the leachate test, nineteen of the twenty tests failed for lead.[18] This was because lead is soluble under both acid and alkaline conditions and the fly ash from modern incinerators is highly alkaline because of the lime used in the scrubbing systems. After the Claremont result, the incinerator industry quickly dropped this argument.

The legal argument used by the incinerator industry was that the ash need not be tested because trash was exempt from testing and ash was only "processed" trash and so that should be exempt as well. This argument went all the way to the US Supreme Court, which ruled in 1994 that if leachate tests showed that the ash was hazardous it had to be sent to a hazardous waste facility.[19] Unfortunately, the court left it to the EPA

FIGURE 4.5. An ash monofill used by the Covanta Energy incinerator located in Haverhill, Massachusetts. *Photograph by Brent Baeslack*

to determine how the testing would be done. At this point the EPA stepped in and gave the house away.

The EPA allowed the bottom ash and the fly ash to be mixed together before testing and a new leachate test (toxicity characteristic leachate procedure test, or TCLP test) made it easier for the combined ash to pass. This was because the lime in the fly ash neutralized most of the fixed quantity of acid used in the test (the previous EP tox test required acid to be added until a specified pH was reached), and the final pH of the leaching solution corresponded to the range of lead's least solubility (lead is soluble at a pH above 10 and below 6, and the final pH with combined ash was about 8 or 9). In other words, not only was there enough lime in the fly ash to protect the fly ash from failing the test, there was also enough to protect the bottom ash from failing as well. If the bottom ash was tested by itself (in other words, without the protection of the lime in the fly ash)—even with the new TCLP test—it would probably fail the test about a third of the time. Because of the much larger quantity involved of bottom ash versus fly ash, this would have been prohibitively expensive for the incinerator industry.[20]

The environmental problem here is that while there was enough lime in the combined ash to neutralize the fixed quantity of acid in the limited test regime, that did not mean that it would do so indefinitely when the ash was placed in a mixed waste landfill producing acid leachate over an extended period of time.

The end result of this EPA giveaway is that today all the combined ash in the United States passes the current leachate test (the TCLP test). Some of this is being sent to huge ash monofills (one being in Haverhill, Massachusetts; see figure 4.5), but much of it is being sent to ordinary municipal landfills—and sometimes is even used as landfill cover.

If handled properly, ash makes incineration prohibitively expensive (especially when the bottom ash is found to be toxic) for all but the wealthiest communities. If handled improperly, it poses both short- and long-term health and environmental dangers.

Argument 6: Incinerators Produce Very Toxic Air Emissions

The modern incinerator burning municipal trash with energy recovery, which came into vogue in the 1970s and 1980s, has left a legacy of unacceptably high levels of dioxins and related compounds in our food, our tissues, our babies, and the environment.[21]

The pollutants incinerators emit include toxic metals,[22] dioxins, and dioxin-related compounds.[23] There are many thousands of these dioxinlike compounds: chlorinated and brominated dioxins, furans, and biphenyls; mixed chlorinated and brominated dioxins, furans, and biphenyls; polybrominated and polychlorinated biphenyl ethers; polybrominated and polychlorinated napthalenes; and sulfur and nitrogen analogues of all of the above. To compound these problems some of these highly persistent or permanent toxins emerge in the form of nanoparticles.

There is no question that since the 1980s the incinerator industry has done a far better job at reducing dioxin emissions from well-designed, well-operated, and well-monitored incinerators—but that does not mean that all incinerators running today, or being proposed in the future, operate (or will operate) in this fashion on a routine basis. There is a world of difference between the theory and practice of incineration, and that difference can ruin the health and well-being of communities that host these facilities, as graphically illustrated in the movie *Trashed*.[24]

The public needs three things to be protected from toxic emissions: 1) strong regulations, 2) scientific monitoring, and 3) tough governmental enforcement of the regulations. If either monitoring or enforcement is weak, the public is not protected by strong regulations. For example, some of the strongest environmental regulations in the world were passed in the former Soviet Union, but that did not prevent horrendous pollution of its environment because these regulations were not enforced.

Timing Is Everything in Local Incinerator Battles

The best chance of beating an incinerator proposal is in the early stages, when the decision is still being discussed by local decision makers. At this point the issue still lies within the democratic process and is open to rejection based on the broader and more basic arguments against incineration. Most incinerator proposals will wither quickly if they are subjected to an objective analysis comparing both the economics and issue of sustainability with the zero waste strategy. However, sometimes decision makers will pass the decision onto the bureaucratic process, saying things like, "We are not experts on environmental and health issues, but the state or the EPA would not permit this to be built if they didn't think it was safe."

This shift is often fatal. Once the decision passes from the democratic to bureaucratic process it is an uphill, costly, and tortuous matter for citizens to combat. Those wishing to build the incinerator will hire expensive engineering consultant companies to prepare environmental impact statements, which may or may not include human health risk assessments. These massive reports—often standing several feet high—are extremely complex for the average citizen to understand, but they invariably end up with a statement that the proposed incinerator poses no threat to man, beast, or insect. Citizens usually will not be able to hire an expensive engineering consultant firm of their own. At this point citizens scramble desperately to find an expert to help them. Those who are prepared to do this pro bono are few and far between.

One expert who has helped many communities in this predicament is Neil Carman from Austin, Texas.[25] Carman is former inspector of the Texas Air Commission and knows the operating problems of incinerators very well. He knows the key difference between the theory and practice of incineration and can usually find places in these proincinerator assessments that are at best optimistic and at worst downright misleading.

When no expert is available, a well-informed public that can ask the right questions is a must. In 2012 when citizens in Arecibo, Puerto Rico, asked for help in dealing with an environmental impact statement prepared for a massive incinerator proposed for their city, Carman prepared a list of questions for them to raise at a public hearing. With Carman's permission I have reproduced this list in appendix A. It's a handy guide for citizens everywhere who find themselves fighting incinerator proposals.

As far as the public is concerned the weak link for incinerators in the United States and many other countries is how infrequently toxic metal and dioxin air emissions are monitored. Incinerator promoters proudly talk about their continuous environmental monitoring (CEM) for their facilities. However,

CEM is not possible for toxic metals (with the possible exception of mercury) and dioxin-related compounds. To monitor these requires inserting a probe into the flue gas and collecting a sample on filters. These filters then have to be sent to a laboratory for analysis, which can take several months. There are not many labs equipped with the very expensive equipment to do this testing or the subsequent analysis. Some countries considering incineration have no companies available to do this testing.

In the United States dioxin testing is only required once or twice a year. As far as adequate science is concerned, this is a sick joke. The operating company gets about one month's notice to get ready for the test. The concern here is that with such a heavy financial commitment at stake, operators may be tempted to manipulate the test results. Such fudging clearly took place in the case of one incinerator in Columbus, Ohio, where plant logs show that workers were told to stock up a separate supply of waste in one of the bunkers—in preparation for the test—that was dry and free of plastic. There is no mistaking the intent of these comments in the plant operator's logs:

> February 15, 1994: "We must have a good source of trash for the test. Hold the north end trash . . ."

> February 21, 1994: "Once again, inform all crane operators to hold north end trash. . . . This test is our future . . ."[26]

Even when the test is done legitimately there are problems. During the test—typically conducted over a twenty-four-hour period—three separate six-hour samples are collected under *stable* operating conditions. In fact, if the operation becomes unstable the testing is stopped. The end result is that what we get is eighteen hours of *ideal* data (or thirty-six hours of ideal data if the test is done at six-month intervals), which is then extrapolated to predict eight thousand hours of *real* operation. Such data do not take into account emissions released during start-up and shutdown, when dioxin emissions are known to greatly increase, nor does it include upset conditions. Of particular concern is that the testing does not include emissions when the company is forced to bypass its air pollution equipment should a blockage occur between the furnace and the stack that prevents the free flow of gas through the system. This point is critical because this equipment is claimed to remove 99.9 percent or more of the dioxin. Thus, during bypass periods dioxin emissions can go up a thousandfold.

The inadequacy of these ideal six-hour tests was emphatically demonstrated by two scientists who compared side-by-side dioxin emissions from the traditional six-hour tests of the flue gas with two-week tests at the same incinerator. After calculations were made, the two-week tests showed that the actual dioxin emissions were thirty to fifty times higher than emissions calculated from the six-hour tests.[27]

This experiment led to the development of a continuous sampling (as opposed to continuous monitoring) system for dioxins, in which two-week filter samples are collected twenty-six times a year so that a whole year's worth of emissions can be calculated. This system is now available commercially (the AMESA system[28]) and is required in Belgium and Germany and is also being used in some Italian incinerators.

It is very suspicious that modern incinerator operators, who claim such great reductions in dioxin emissions, are not eager to use this continuous sampling system to prove their case. Some incinerator promoters claim that the AMESA system is too expensive, but in my view it is a small price to pay to win the confidence of the public.

The poor science of the six-hour tests is further compounded by poor statistical analysis of the data collected. Typically, operators report an *average* of the three measurements (the three separate six-hour tests) when what should be reported is a 95 percent upper confidence interval. This is a statistical analysis based upon the variation in the data. This analysis predicts what the level would be *below* 95 percent of the time if the test was repeated 100 times under identical physical conditions. This is the kind of number that would be meaningful for input into health risk assessments and for extrapolation from the eighteen hours of data to a full year's operation. The bottom line here is that if there is any significant variation in the three tests the upper confidence level will not only be higher than the average but also higher than the highest of the three tests!

The very fact that most of the dioxin is removed in the pollution control devices, when they are operating properly, means that the dioxin ends up in the ash. This is not being measured and is conveniently ignored in national dioxin inventories carried out by agencies like the EPA.

Meanwhile, as dioxin emissions have decreased, the toxicological concerns over dioxin and dioxin-related compounds have increased. However, there has been determined effort by several industries to hide those increased toxicological concerns from the public.

Incredibly, the publication of the EPA's reassessment of dioxin's toxicity (urged on them by the paper industry in 1992) was essentially completed in

1994—but the final report was successfully delayed by industrial and agricultural interests for over eighteen years. In 2012, the risk assessment of dioxin's *noncarcinogenic* effects was finally published.[29] Two commentaries on this release can be accessed on the Internet.[30, 31] At this writing, though, we are still waiting for the final draft of dioxin's carcinogenic effects!

The first warning about dioxin interfering with human hormonal systems—even at background levels—came in 1992 from six Dutch researchers who published a short article in the form of a letter to the *Lancet*.[32] They reported that one-week-old babies born to mothers with a high versus a low background level of exposure to dioxin, based upon the level of dioxin in their breast milk, had a significant difference in thyroid hormone levels and ratios.

Since then dioxinlike compounds have been shown to interfere with six different hormonal systems (male and female sex hormones, thyroid hormones, insulin, gastrin, and glucocorticoid).[33] Thus, dioxins not only have the potential to interfere with sexual development but also mental development and the immune system.[34]

While the EPA dragged its feet for eighteen years on releasing its final health risk assessment, the US Institute of Medicine (IOM) published a report in 2003 suggesting ways that people—especially young girls—could limit their exposure to dioxinlike compounds. The report stated that:

> Fetuses and breastfeeding infants may be at particular risk from exposure to dioxin like compounds (DLCs) due to their potential to cause adverse neurodevelopmental, neurobehavioral, and immune system effects in developing systems.

The IOM panel recommended that "the government place a high public health priority on reducing DLC intakes by girls and young women in the years well before pregnancy is likely to occur" by encouraging them to choose foods lower in animal fat and substitute low-fat or skim milk for whole milk.[35] Those who are so proud of incineration should face up to the fact that the legacy of their recent history is we are having to interfere with the diet of a huge part of our population in order to avoid the dioxin accumulated in the food supply by the incinerator industry's past emissions. We need to be wary about the promises from a new generation of incinerator promoters who are boldly claiming the incineration dioxin problem has been solved.[36]

Waste reduction, reuse, recycling, composting, and redesign provide a far less drastic way to fight off the dioxin (and other pollution) threat of incineration

The Dioxin Threat in a Nutshell

Dioxin concentrates in the fat of animals and fish.[37]

Once consumed by humans the most toxic dioxins are only very slowly metabolized by the liver, and as a result steadily accumulate in our body fat. The male has no way of getting rid of this dioxin, but a woman does: by having a baby. During the nine months of pregnancy the dioxin in the woman's body fat (accumulated over twenty years or so) moves into the fetus. The resulting concentration and toxicological impact is thus hugely increased.

The fetus and the breast-fed infant get the highest doses of dioxin and related compounds. This is very serious because both fetal and infant development is under hormonal control, and dioxins and dioxinlike compounds interfere in a variety of ways with hormonal systems.[38]

than attempting to change the diet of millions of our citizens. But unless we are simply to revert to more use of landfills, we need an effective combination of all these listed options in order to offer both an alternative to incinerators *and* landfills. That is what zero waste sets out to do.

Argument 7: Incinerators Release Very Toxic Nanoparticles

Concern over nanoparticle emissions from incinerators has grown significantly since 2000. Nanoparticles, sometimes referred to as ultrafine particles, are particles of less than one micron in diameter. Nanoparticles are not new. They have been with us since the advent of high-temperature combustion, which includes gas and diesel engines; coal-, oil-, and natural-gas-fired power stations; and even wood burners. What is new is the birth of a discipline called nanotoxicology. This new area of research was brought about by the advent of nanotechnology in which nanoparticles have been added to medicines and commercial products ranging from shaving cream to tennis rackets. This new technology has spawned the question: Is there anything potentially dangerous to health from nanoparticles?

The answer is in. Yes, there are very serious health concerns.

The key concern is that these very tiny particles can easily cross cell membranes. Thus the normal defense mechanisms preventing particle entry to tissues do not prevail with nanoparticles. Nanoparticles in the air we breathe can cross the lung membrane (see figure 4.6), and the nanoparticles in our food can cross the membranes of the GI tract.

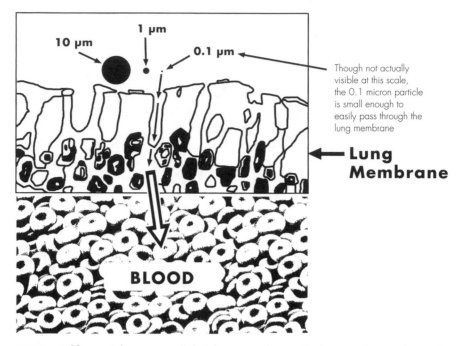

FIGURE 4.6. Nanoparticles are so small that they can easily cross the lung membrane and enter the bloodstream. *Illustration by Jack Cook*

In both cases the nanoparticles can get into the bloodstream. They can then travel to every tissue in the body and enter these as well. They can even cross the blood-brain barrier.[39]

Of all the high-temperature combustion sources, the nanoparticles from trash incineration are the most worrying because they are the most toxic. Incinerators take in essentially all the toxic elements we use in the manufacture of products. These include the pigments or stabilizers in plastics; the toxic metals used in batteries and electronic equipment; and chlorinated, fluorinated, or brominated compounds used in plastics and fire retardants.

Worse yet, nanoparticle emissions are neither being regulated nor monitored in incinerators. The particle sizes regulated in incinerator emissions are generally 10 microns; in some countries this may be going down to 2.5 microns. Either way, comparing particles of this size to nanoparticles is like comparing cannon balls to grains of sand (see figure 4.7).

There has been a surprising lack of response to concerns expressed on the nanoparticle issue from the incinerator builders and the governments that promote them. For example, I have seen no science-based industrial or governmental response to this problem—a point also made in the journal

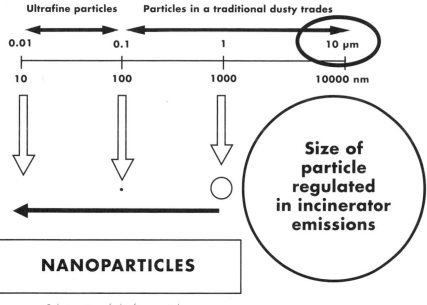

Relative size of ultrafine particles
compared with particles in traditional dusty trades

FIGURE 4.7. Nanoparticles are much smaller than the size of particles currently regulated in incinerators. *Illustration by Jack Cook*

Environmental Health Perspectives (published by the US National Institute for Environmental Health Sciences [NIEHS]) in 2006,[40] and I have only seen simplistic and dismissive responses to the widely distributed thirty-page presentation on this issue by Professor Vyvyan Howard (who had previously coedited a book on ultrafine particles in 1999[41]) before a hearing board on a proposed incinerator for Ringsakiddy, Ireland.[42]

It is already well established that disease rates in large cities can be related to air particulate levels. Both morbidity (from respiratory problems and heart disease) and mortality rates go up as the level of particulates go up.[43] Moreover, as the particle size measured goes down this relationship gets stronger.[44] It is only logical that the situation will get worse if more nanoparticles are added to the mix. It stands to reason that if an incinerator is built in an already polluted city, both morbidity and mortality rates are going to increase because of the very toxic nanoparticles they inevitably release. In other words, it will make a bad situation in a polluted city even worse.

Those who promote incineration have a slick (sick?) public relations way of handling these realities. They have couched this discussion in a term: deaths brought forward (DBF). Here is a how one incinerator company describes it:

> It is important to note that an increased risk of mortality and morbidity, due to elevated PM10 (particulate) exposure, is small and limited to a fraction of a population, which is already in poor health. As such the term DBF does not constitute new/additional deaths but a reduction in life expectancy for those whose health is already seriously compromised.[45]

What they are saying here is that incinerator emissions don't kill people, they simply shorten their lives! Such a perverse sentiment brings the public relations of incineration promotion to a new low level.

Nanoparticles are so small that they evade most air pollution control devices. The key question for the incineration industry is, "How does a modern incinerator propose to capture these particles?" Or are they simply prepared to duck this question because the nanoparticles are neither regulated nor monitored? If they have no clear and concrete answers to this key question surely this issue must trigger the precautionary principle[46]—especially when there is now an alternative strategy available that can achieve comparable (and even greater) reductions in the amount of trash landfilled.

Argument 8: Incinerators Are Extremely Unpopular with the Public

Because of many of the concerns above, incinerators are hugely unpopular with the public. When citizens probe beneath the smooth veneer of promotion, they see that what at first might appear to be a quick-fix solution does not look so quick once the political and legal battles begin. Again, let's remember that more than three hundred of the roughly four hundred incinerators proposed for North America between 1985 and 1995 were defeated.[47] There were reasons for that, and many of those reasons remain unchanged.

Argument 9: Incinerators Are Not Sustainable

Incineration of municipal waste dates back to the nineteenth century (the first WTE plant was operating in Hamburg, Germany, in 1895).[48] Incineration has no place in the twenty-first century. On the global scale, incinerators waste energy and waste the opportunity to really fight global warming. On the local scale, incineration wastes the opportunity to create jobs and otherwise help the local economy in a sustainable fashion.

Even if the finest engineers could solve the nanoparticle and ash disposal problems, they would not make the incineration of trash acceptable. It simply doesn't make either ethical or economic sense to spend so much time, money, and effort destroying materials we should be sharing with the future. Those who have been preoccupied with making incineration safe have lavished their engineering ingenuity on the wrong question. Society's task today is not to perfect the destruction of our waste but to stop making products and packaging that have to be destroyed.

Our future will be dominated by the need to find sustainable ways of living on the planet. When a community builds an incinerator, it is advertising to the world that it was not clever enough politically or technically to recover its discarded resources in a manner moving toward sustainability. In other words, we waste more with incineration than material resources.

Argument 10: There Are Better Alternatives
And that is what the rest of this book is about.

The Early History of Zero Waste

In the early 1990s a group was set up in Manila that called itself the Zero Waste Recycling Movement of the Philippines. This little group—led by Luz Sabas—was remarkably creative in the ways it found uses for every scrap in the waste stream. Roof tile was made from tin cans, curtains were made from drinking straws, compost kits were made from car tires, and even a pesticide was made from cigarette butts. However, few people outside the Philippines heard about their efforts. Even among other environmental groups in the Philippines their message gained little traction as it was not seen as a practical way of convincing government officials not to build incinerators. The major push for the notion for zero waste as a serious alternative to landfills and incinerators had to wait for developments elsewhere.

The Convergence of Two Ideas

In his book *The Ghost in the Machine* the late author Arthur Koestler discusses a common feature of the creative and discovery processes whether in humor, art, or science. He coined a term called "bissociation." By this he meant the intersection or collision between similar ideas from two different frames of reference.[1]

So what were the two ideas that came together to produce "zero waste"? In this case the two ideas were very similar but were being developed in different frames of reference and in different places—in fact as far away from one another as you can get on our planet.

In the 1980s in Berkeley, California, Daniel Knapp started a small operation scavenging discards from the landfill. By the 1990s this had developed into a major reuse and recycling operation, Urban Ore. In the process Knapp came up with the notion that *all* discards could be divided up into twelve major categories. This further evolved into purposely designed ecoparks where the twelve categories could be efficiently collected, processed, marketed, or used on-site. Based on these discard categories he argued for "total or 100 percent recycling." Many thought that this was an impossible goal and that it was totally unrealistic to talk about it.

Meanwhile, at the other end of the world the leaders of Australia's capital city Canberra were meeting with citizens to discuss how to minimize the flow of waste to their landfill. When the politicians asked the citizens how much waste they felt should be going into the landfill, they said, "None!" Such a position was based more on a feeling of moral obligation than a realistic analysis of what was achievable. Remarkably, however, the politicians accepted the challenge and thus was born what eventually became the No Waste by 2010 law, which was passed by the Australian Capital Territory (ACT) government in 1996.

But the concepts of "total recycling" and "no waste by 2010" began to merge into "zero waste" a year before this landmark legislation, when Knapp was invited to Australia to share his ideas. There he presented slide shows from dawn to dusk in Canberra and Melbourne on both reuse and on "total recycling," explaining his twelve master categories and sharing drawings of ecoparks designed to handle all of these twelve categories. He was also taken to look at a reuse operation, called Revolve, at a landfill. Like Knapp, his hosts saw the merits of scavenging waste, sorting it, and reusing it. The meeting was an important verification for both Knapp and the Revolve workers that they were on the right track.

Knapp was a master of reducing and reusing waste, but the idea of getting to *no waste* in a short time frame was intriguing for more than one reason. "The thing that fascinated me at the time," he says, "was that the zero waste idea, even though first enunciated by citizens acting through Revolve, was being most effectively championed by a bunch of bureaucrats on the eighth floor of a government building in Canberra. It was they who drafted the legislation No Waste by 2010 that went to Parliament and was passed in 1996." Those officials had even hired an engineering consultant to draft a plan for an impressive-looking resource recovery park. Revolve, too, had already drafted a plan—a much more practical design based on their idea of maximizing the recovery of discarded resources.

Knapp and the Australians exchanged copies of their concepts and site plans, and Knapp came home and shared them with a leading landfill opponent—activist Bill Sheehan. Sheehan had recently helped to set up the GrassRoots Recycling Network (GRRN). He thought the fact that the ACT government was backing zero waste was big news and put out word over the Internet. The message that Australians were endorsing zero waste went viral back in the United States, and the concept gained popularity. Says Knapp, "I found it was a lot easier to sell 'zero waste' than 'total recycling.'"

Back in California, Knapp and others began to make tremendous strides with zero waste. Ironically, though, the news from Australia took a turn for the worse. Knapp continued to make trips to Australia to collaborate. But he recalls, "All the people I had been talking with in Canberra up until about 2000 were swept out of power when a new party took control of government. What makes me most upset about this is that Revolve—the plucky little nonprofit refuse operation closest to Urban Ore in structure of any company I ever visited—was driven into bankruptcy."

Knapp remains puzzled to this day why his idea of "total recycling" did not catch on—even with activists—while "zero waste" was embraced by all and sundry. So what was the difference? Why was one message largely ignored while the other spread rapidly through California and beyond? Knapp's story highlights two reasons: the fact that the ACT *government* endorsed zero waste and passed a law to accomplish it, which suddenly brought the matter from the realm of fantasy into reality; and the fact that the idea was raised on the Internet, which by the mid-1990s was gaining considerable traction among grassroots activists as both an educational and activist tool.

But there are other factors, too. Total recycling implies that the community has to take on the whole waste problem by themselves; the responsibility to recycle everything lies with them. On the other hand the zero waste philosophy recognizes that communities cannot get rid of waste alone. They need industrial help in the form of industrial responsibility. We have to redesign the things that we currently can't reuse, recycle, or compost out of the manufacturing system.

Timing was also critical. The ground was particularly fertile for the zero waste message to spread in California in the mid-1990s. If it had reached there a decade before, when plans to build thirty-five incinerators were being debated, I think the idea would have been dismissed out of hand. By the same token, if the message had been brought to Canton, New York, in 1985—where I was residing and had begun the campaign to fight a local incinerator proposal—the idea would also have been laughed out of court. I vividly remember one member of the county solid waste team shouting down the phone at me in 1985, "Paul, if you talk about more than 15 percent recycling you are going to lose all your credibility!"

But by 1995, California had only built three of the thirty-five proposed incinerators and there was no way that the public would accept any more. Meanwhile, landfills were being sited further and further away from urban centers, with transport costs and tipping fees rising.

In 1989, the California legislature had passed a law requiring all communities to divert 25 percent of waste from landfills by 1995 and 50 percent by

2000. By 1995, many communities had achieved this goal and were saving money in the process. Both citizens and decision makers were asking why they should stop at 50 percent—why not even higher? Thus, there were many people in California—including decision makers—primed to accept a zero waste message.

Zero Waste Fits Well with Many Other Trends

The zero waste idea also fit well with the evolution of many of the following trends and realizations, which were peaking in the 1990s.

The Rise of Recycling

Although there has always been entrepreneurial recycling of the valuable items like scrap metals, in the 1970s recycling of other materials, like newspapers, was seen as something for the Boy Scouts of America and other voluntary groups to do. The enemy of commercial recycling of the less valuable materials was cheap landfilling, which in the 1970s was widespread. With the advent of less political space for new landfills, the tipping fees began to rise. With the introduction of expensive incinerators in the 1980s, tipping fees rose even higher. At this point the notion of avoided disposal fees entered the picture. Decision makers realized it was cheaper to recycle than to landfill and much cheaper than incineration, even if one received little or no income from the sale of the material. The same economic arguments drove yard waste composting. This was seen as the cheapest disposal option for a sizeable fraction of the waste stream. Similarly, recycling of construction and demolition debris became very attractive because construction debris is so heavy—and thus so expensive to haul away and discard. With the advent of curbside collection and blue box programs it also became clear that despite the earlier pessimism of decision makers, North Americans would recycle if the programs were made simple and convenient. In fact, Neil Seldman of the Institute for Local Self-Reliance would tell audiences in the early 1990s, "More people recycle in the US than vote!"

The Development of Community Reuse Centers

There have always been operations involving reuse of discarded materials in the United States and other countries. These range across a whole spectrum of yard sales, garage sales, church jumble sales (in the United Kingdom), pawn shops, antique shops, and large charity thrift shops. Some of these, like the Salvation Army and Goodwill Industries International, have also consciously built into their operations employment opportunities in both sales and repair. Operations like Urban Ore; ReSOURCE North; Wastewise in Georgetown, Ontario; and Revolve in Canberra have taken these up a notch to involve the whole community and have linked themselves with other businesses like those deconstructing old buildings. The end result is that, like recycling, reuse is taking an ever-larger chunk of the discard stream and making money and creating jobs doing it. These facilities also make the benefits of both recycling and reuse very visible to the average citizen—far more so than having one's recyclables removed from the curb.

The Growth of Composting

I have always felt that the bigger threat to incineration was not recycling but composting. Composting can contribute as much as half the diversion rate from landfill, especially in warmer climates with rapidly growing vegetation. Moreover, once the "smelly" organics are taken out of the waste stream, the recyclables are easier to store and handle. Since the 1980s we have seen composting transition from backyard composting to curbside collection of yard waste to curbside collection of food waste.

The Need to Fight Overconsumption

Since the 1950s there has also been a growing unease over the excesses of consumption involved in the throwaway society. Vance Packard, author of the book *The Waste Makers,* was one of the early writers drawing attention to the techniques of advertisers to encourage us to buy more than we need as well as the evils of built-in obsolescence.[2] A series of major TV programs and documentaries have drawn attention to our wasteful ways;[3] many of these are listed in the Resources section at the end of this book. Images of overflowing landfills and wildlife choking on our debris have drawn particular attention to many throwaway plastic items. More recently the work of Charles Moore has drawn attention to the sickening plastic pollution of the Pacific Ocean;[4] his work is featured in the movie *Trashed.*[5] As a result of all this publicity a great deal of wrath has been directed toward the manufacturers and users of Styrofoam cups, plastic bottles, and plastic bags—leading to bans on these items in many jurisdictions. This discontent with the throwaway society feeds smoothly into the desire for a zero waste society—one in which industry is forced to behave more responsibly. More and more jurisdictions are demanding EPR from manufacturers, which forces attention on the front-end of the problem, a key driver of the zero waste strategy.

The Growing Awareness of the Insidious Toxic Invasion of Our Lives

Between 1985 and 1995, more and more people also came to appreciate the insidious dangers of many of the pollutants released by incinerators (and more recently disposable plastic packaging).

From the days when Rachel Carson's *Silent Spring* (1962)[6] exposed the hazards of DDT continuing through more recently when Theo Colborn's *Our Stolen Future* (1994)[7] cast public attention on the way industrial toxins can enter our bodies and disrupt our endocrine systems, the evidence has mounted that industrial chemicals once thought to be safe can build up in our bodies and cause harm to future generations. Indeed, this harm is often not apparent at the time of exposure and may take many years—as many as twenty or forty—to manifest itself. This is particularly the case for endocrine-disrupting chemicals like dioxins and furans.

In short, zero waste marries the growing public concerns about both toxicity and sustainability.

PART TWO

Zero Waste Initiatives
Around the World

In the following eight chapters we look at zero waste initiatives in different regions in the world. While the various programs in the pages ahead take place in communities of various sizes and have different features, they all share a couple of things in common.

First, zero waste success stories begin with individuals—individuals with vision and determination. They come from all walks of life: academia, the community, business, nongovernmental organizations, and local government.

Second, these stories underscore the truism that communities recycle, not countries. While national and state or regional governments can pass helpful legislation—and sometimes unhelpful legislation—the key issue for zero waste advocates to focus on is finding what your local community can do for itself. The process is best started by exploring what other communities of a similar size and demography have been able to achieve—and also what obstacles they have faced.

Not surprisingly, the news in these chapters is not all good. In addition to many successful models, there are examples of programs that started with high hopes but have faltered (as in Canberra, Australia). There are other examples where the lofty rhetoric of zero waste has been corrupted by proposals to build incinerators (as in Vancouver, British Columbia). There are also examples of other programs failing to take a pioneering role in the development of a genuine zero waste strategy when they had a golden opportunity to do so (as in Edmonton, Alberta).

Troublesome, too, is that some of the globe's most densely populated countries are heading in the wrong direction on the waste front. With China threatening to double its incinerators and India announcing plans to build a third incinerator in New Delhi, it looks as if these two huge countries with their booming economies are ignoring the lessons learned in Europe and North America. Moreover, their rush to build capital-intensive incinerators threatens the livelihood of millions of ragpickers, who know more about zero waste from firsthand experience than any other people on earth. Fortunately, such projects are not moving forward without strong resistance from citizens' groups.

Currently, the most important part of the struggle for zero waste in the Global South is the challenge to local governments to incorporate the efforts of ragpickers into their waste management plans and to provide them with better and safer working conditions—as well as health care and educational opportunities for their children. For this to happen the ragpickers need to be organized into cooperatives that can negotiate free and safe access to discarded materials. More than anything else, these people need our respect—not our disdain. South America offers some very hopeful examples of inclusion for these informal workers. Sadly, Egypt offers a far more distressing example of how the good work of ragpickers and recyclers has been decimated by clumsy waste management planning by the municipal government.

Hopefully, even at this late hour, India and China can be persuaded to copy the world's best examples—not its worst.

Today, those examples can be found in California, Italy, and Spain.

The United States

As a whole, the United States is far from a forerunner when it comes to zero waste initiatives. The nation has so much space that even when communities organize to stop incineration[1] it does not signal that they will embark on innovative waste-handling programs. Some have embraced recycling programs, but far too many other communities have settled for limited recycling with the bulk of their waste being exported to distant landfills. The big waste companies still dominate the agenda.

Consequently, on average only about 34 percent of American waste is diverted from landfills with recycling, composting, or other reuse initiatives (40 percent if construction and demolition debris is included).[2] However, there is one state, California, that leads the pack on waste reduction—and one city in that state, San Francisco, is a recognized world leader of large cities on the zero waste frontier.

As new incinerator proposals make their appearance and communities organize to stop them, California offers practical models (as opposed to theoretical appeals) to which grassroots activists can point. We can thus expect more communities in the United States to adopt the zero waste strategy in the future—some because it is the right (sustainable) thing to do and some because they are threatened by new incinerator proposals and they need a practical alternative to offer their decision makers.

California

It makes sense, then, to start our review of zero waste initiatives worldwide with a review of developments in California.

In the 1970s, the rules on handling waste were in a state of flux as California moved from open burning dumps to sanitary landfills. In 1972, two key events took place.

First, the state passed the Solid Waste Management and Resource Recovery Act, which created the California Solid Waste Management Board (CSWMB). It also called for county waste plans, public involvement, and local enforcement agencies. Second, the California Resource Recovery Association (CRRA) was formed, which was the nation's first statewide recycling organization. This organization played an important role in formulating the waste agenda in California in the 1970s and has continued to do so ever since.

The CRRA has now become one of the largest recycling organizations in North America. Its annual convention attracts hundreds of participants from local government, the waste industry, recycling businesses, grassroots organizations, and educational bodies.

In 1976, the CRRA supported a resource-recovery antilitter bill being pushed by the packaging industry to fund recycling programs, WTE research and development, and antilitter programs in all counties.

The good thing about this law was that it established local or state funding for all serious nonprofit and public recycling programs. As a result during the 1970s and 1980s local programs grew and cities began to institutionalize curbside recycling programs as well as statewide litter-control programs. The bad thing was that millions of dollars were spent on WTE (incinerator) planning throughout the state. But public opposition prevailed; of the thirty-five trash incinerators proposed by 1995, only three were built.[3]

Meanwhile, the CRRA was holding several conferences that planned for a more progressive future as far as recycling was concerned. This effort culminated in the drafting of a state recycling agenda that eventually led to the California legislature passing the California Integrated Waste Management Law in 1989[4]—not only requiring all communities to reach a diversion of 25 percent from landfill by 1995 and 50 percent by 2000 but also establishing a fine of $10,000 a day for communities that did not reach these deadlines. Unfortunately, the law also allowed diversion credits for burning.

The California Integrated Waste Management Board (CIWMB) was to oversee the whole program. This board established what's called a base-year solid waste generation figure for each community. In other words, they calculated the average amount of waste each community normally sent to landfills in order to determine the percentage of solid waste diverted each year. To help communities achieve the diversion goals, "each jurisdiction was required to create an Integrated Waste Management Plan that looked at recycling programs, purchasing of recycled products, composting and waste minimization."[5]

By 1996, nearly three hundred communities in California, both large and small, had achieved a 50 percent diversion rate. Many found that this was actually saving them money by cutting out both the long transport costs to landfills and the tipping fees charged. As a result of this successful reduction some began to wonder whether they should go further than the 50 percent state-required goal—a sentiment amplified by news of Canberra's No Waste by 2010 law. This news from Australia triggered a number of more progressive communities to discuss adopting a zero waste program. Among the first to do so were Del Norte County (the northernmost county in California) and San Luis Obispo County. The CRRA responded to these new initiatives by drafting the Agenda for a New Millennium, which called for the adoption of zero waste statewide.

San Francisco: The Flagship for Zero Waste in a Large City

By 1999, San Francisco had adopted a program with an interim goal of 75 percent diversion by 2010 and an ultimate goal of zero waste by 2020—ambitious and impressive for a city with a population of about 850,000, very little space, and where educational programs have to be carried out in at least three languages (English, Spanish, and Chinese). For me, its 2010 goal of 75 percent diversion was even more exciting than its zero waste goal by 2020, because some of the decision makers who committed to that goal knew that they would still be around in 2010 and thus would be held politically accountable if they failed to deliver. But deliver they did.

By early October 2012 San Francisco had reached 80 percent diversion from landfill, and they did this without burning any of the residual fraction.[6] There are several features of this program that have made it successful.

First, San Francisco was the first large jurisdiction in the United States to collect kitchen waste in addition to yard waste (many collect the latter). San Francisco did this using a color-coded three-bin system—one bin for compostables (green), one for recyclables (blue), and one for residuals (black) (see figure 6.1). These are picked up once a week.

Second, San Francisco officials knew that to get compost of sufficient quality to give local farmers the confidence to use it to grow food or put on their vineyards they would have to collect very clean material.

To ensure this they reduced collection fees by 25 percent for restaurants and hotels putting out separated organic waste instead of mixed waste. The city also sent municipal employees to restaurants and hotels to educate kitchen staff on exactly which materials could go into the green

Milestones on San Francisco's Journey to Zero Waste

50 percent waste diverted by 2000
63 percent waste diverted by 2004
70 percent waste diverted by 2008
72 percent waste diverted by 2009
75 percent waste diverted by 2010
78 percent waste diverted by 2011
80 percent waste diverted by 2012

FIGURE 6.1. San Francisco's three-container collection system. *Photograph by SFEnvironmental*

FIGURE 6.2. Kitchen staff gather around the organics collection container destined to produce higher-quality compost for farmers and wine makers. *Photograph by Larry Strong, courtesy of Recology*

FIGURE 6.3. San Francisco buses carry a procomposting message. *Photograph by SFEnvironmental*

FIGURE 6.4. In this street booth an employee of SFEnvironmental explains to a passerby exactly what goes into each container of the three-bin system. *Photograph by SFEnvironmental*

FIGURE 6.5. A display at a farmers' market illustrates what materials go into each container. *Photograph by SFEnvironmental*

containers. Kitchen staff in both hotels and restaurants are proud to be part of this important program (see figure 6.2).

Third, San Francisco officials used a very aggressive advertising campaign to get news of the program out to citizens. Every bus shelter sported an announcement telling people that "recycling is now as easy as 1-2-3!" and the buses themselves carried ads promoting the need to separate organics (see figure 6.3).

These public advertising campaigns were also buttressed with widely distributed leaflets to homeowners, booths on the streets to explain the system as well poster displays in public places that clearly illustrated what was supposed to go into each container (see figures 6.4 and 6.5).

FIGURE 6.6. *Top:* Recology picks up the recyclable (blue container) and residual (black container) fractions in dual-chamber trucks (and a separate truck for the green container). Note that the side of the truck has been painted with a 3D image to show the inside of the truck to make this operation transparent to observers. *Bottom:* A close-up of the painting. *Photographs by Larry Strong, courtesy of Recology*

Also adding to the effectiveness of the program was the fact that the company (now called Recology) that won the contract for the pickup of these containers owned the recycling and composting facilities but not the landfill. Thus, it had no incentive to send waste to its competitor that ran the landfill, but it had a huge incentive to maximize recycling and composting. Finally, the city introduced a bonus scheme to Recology in the event that higher diversion rates than the annual targets were achieved.

With the introduction of the city's mandatory recycling and composting ordinance in 2009, incentives in the collection service cost structure meant that businesses could save up to 95 percent of service costs by participating in recycling and composting programs.

The Recology vehicles used to pick up the blue and black containers are shown in figure 6.6. Note the 3-D painting the side of the vehicles. Recology spokesperson Robert Reed explains the significance of this: "Every other garbage company on the planet is trying to hide garbage. You go to Atlanta, Chicago, etc. You put refuse in a trash bin. You never see it again. In San Francisco we put 3-D photos of garbage on the sides of our trucks. It is as though you can see inside the truck. It is an effort to get people to look at their garbage. When people do, they see that garbage is not garbage at all. It is a mix of resources that should be recycled and composted."[7]

COMPOSTING

According to GAIA, since this ordinance was passed there has been a "50 percent increase in businesses using the compost collection service and a 300

FIGURE 6.7. San Francisco's composting facility is surrounded by farmland, far enough from residential populations to avoid odor issues. *Photograph by Larry Strong, courtesy of Recology*

percent increase in the number of apartments using the service. As a result, the collection of compostable materials has increased by 45 percent so that nearly six hundred tons per day of food scraps, soiled paper, and yard trimmings are sent to composting facilities."[8]

San Francisco's composting facility is located approximately seventy miles from the city (see figure 6.7), far enough away from heavily populated areas to avoid odor issues and closer to farms that can use its product.

Local farmers are glad to use the compost in the San Francisco facility to grow fruit and vegetables—as well as in their vineyards (see figure 6.8). All these products go back to San Francisco, thereby helping to close the loop between food production and discarded food scraps.

FIGURE 6.8. Local farmers use the compost to produce fruit, vegetables, and wine, as in this vineyard. Some of the produce and products go back to San Francisco. *Photograph by Larry Strong, courtesy of Recology*

FIGURE 6.9. The cover crop emerging between rows of vines is grown with compost produced from San Francisco's organic discards. *Photograph by Larry Strong, courtesy of Recology*

FIGURE 6.10. Here we see an area where cover crop has been applied, along-side one where it has not (bottom). *Photograph by Larry Strong, courtesy of Recology*

Says Recology's Robert Reed, "San Francisco's composting program is combining something old—growing cover crops to fix nutrients and carbon in the soil—with something new—urban compost collection programs. The program is a hybrid and represents a partial solution to our environmental challenges. Agronomists say if every city replicated San Francisco's urban compost collection program, we could offset more than 20 percent of the nation's carbon emissions."[9]

JOBS, JOBS, AND JOBS!

The San Francisco program also highlights an extremely important aspect of the zero waste strategy: the ability to create a large number of jobs compared to either landfilling or incineration. San Francisco's major recycling plant (Material Recovery Facility [MRF]) is located on Pier 96 in San Francisco harbor (see figure 6.11) and is owned by Recology.

Recology employs over one thousand unionized workers represented by the Teamsters. According to GAIA, "Recology is a private, employee-owned company—so there are no corporate bosses, no stockholders and no possibility of a hostile takeover by the waste industry."[10]

Mayor Gavin Newsom put it this way: "San Francisco is showing once again that doing good for our environment also means doing right by our economy and local job creation. . . . For a growing number of people, recycling provides the dignity of a paycheck in tough economic times. The recycling industry trains and employs men and women in local environmental work that can't be outsourced and sent overseas, creating ten times as many jobs as sending material to landfills"[11] (see figures 6.12 and 6.13).

FIGURE 6.11. San Francisco's largest recycling facility, located on Pier 96
Photograph by Larry Strong, courtesy of Recology

FIGURE 6.12. Unionized workers in the Pier 96 facility separating different grades of paper. *Photograph by Larry Strong, courtesy of Recology*

FIGURE 6.13. Workers separating bottles, cans, and other containers. *Photograph by Larry Strong, courtesy of Recology*

MORE WAYS TO REDUCE WASTE

In 2006, San Francisco introduced the Construction and Demolition Debris Recovery Program, which made recycling C&D debris mandatory; as a result San Francisco now sends 20 percent fewer tons of this material to landfill.[12] Also in 2006, the city passed the Food Service Waste Reduction Ordinance, which bans polystyrene foam take-out containers and requires containers to be recyclable or compostable. Almost all of San Francisco's restaurants are participating in this program.[13]

The city also focuses on education in the form of free training for businesses and apartment buildings. According to GAIA, "In 2011, the city worked directly with approximately 300 apartment buildings (encompassing 21,000 units), 800 commercial accounts, 4,000 food establishments, and over 100 of the largest events."[14] In addition the city has trained over four thousand of its own municipal employees to ensure recycling and composting takes place in its own buildings. This alone has saved the city over half a million dollars in city trash service fees and other efficiencies.[15]

The San Francisco program has attracted attention around the world. It was the solid advances made by a city of the size and complexity of San Francisco that encouraged the mayor and vice mayor of the city of Naples to remove the city's support for a proposed incinerator and instead adopt a zero waste strategy (see chapter 7). When the citizens of Parma, Italy, were faced with the prospect of an incinerator close to their world-famous food industry, they mounted a massive campaign to stop the project. They also asked San Francisco and its mayor for support.

Mayor Gavin Newsom responded by sending a letter to the mayor of Parma expressing his dismay at Parma's incinerator project.[16] Then Parma pharmacist Francesco Barbieri traveled to San Francisco with professional videographer Vincent Ivarez to see and capture the city's waste-handling operations for folks back home.[17] This videotape in turn has inspired other communities in Italy to declare a zero waste strategy.

Other Zero Waste Programs in California

Several other major cities in California have also embarked on a zero waste program—including Alameda, Berkeley, Chula Vista, El Cajon, Los Angeles, Oakland, Oceanside, and San Diego. Several counties have also declared zero waste programs—these include Alameda, Del Norte, Monterey, and San Luis Obispo.

At the statewide level the legislature has recently upped the ante by passing a law (AB 341) that raises the required diversion goal for all communities in the state to 75 percent by 2020. This is to be achieved by source reduction, recycling, and composting.[18] There are also several waste-related companies and other corporations that have made a tremendous difference for the state's zero waste goals.

URBAN ORE IN BERKELEY

If some of the best environmental things happen first in California, then we can also say that in California the best things usually happen first in Berkeley (e.g., Berkeley was the first community to ban Styrofoam packaging). In addition to a very long and successful recycling program, Berkeley is the home of one of the most important reuse and repair operations in the world—Urban Ore, discussed in chapters 2 and 15. Urban Ore has set the standard for recovering reusable objects from the discard stream as well as creating a large number of well-paid jobs doing it. I videotaped the facility in 1999.[19]

ALBERTSONS

Another business that is pioneering a zero waste strategy in California is the grocery store Albertsons. Two of its stores in Santa Barbara received the 2011 Governor's Environmental and Economic Leadership Award (GEELA) for their zero waste efforts—namely, for recycling, reusing, or composting 95 percent of waste products from both stores.[20] Together these two stores are diverting over 2 million pounds of noncontaminated waste from the landfill annually. This includes over 800,000 pounds of cardboard, over 27,000 pounds of plastic and over 2,600 pounds of paper. Additionally, they are donating on average almost 150,000 pounds per year of products to the Foodbank of Santa Barbara County, the equivalent of 125,000 meals.[21]

Meanwhile, Supervalu (Albertsons' parent company) has committed to transitioning a total of forty stores to zero waste in 2012. Albertsons expects to help drive an even more aggressive goal for the company in 2013. According to the company, their initiatives "are good for the environment, good for its communities and good for its overall business."[22]

These successful initiatives by Albertsons have huge ramifications for many other food stores in California and beyond. Albertsons currently operates 447 supermarkets in southern California, southern Nevada, Idaho, Montana, North Dakota, Oregon, Utah, Washington, and Wyoming. Its parent company, Supervalu, is one of the largest companies in the US grocery category, with estimated annual sales of $38 billion. It operates a network of approximately 4,270 stores across the United States with approximately 154,000 employees.[23]

Again—good environmental things often happen first in California and then spread across the country. Several other corporations there have taken on zero waste goals, including Anheuser-Busch, Apple, Fetzer Vineyards, Hewlett-Packard, Mad River Brewing Company, Ricoh Electronics, and Toyota.

Boulder, Colorado

Boulder plays hosts to Eco-Cycle (www.ecocycle.org), which is one of the oldest and largest nonprofit recyclers in the United States. It was created in 1976 by everyday residents who had a passionate belief in conserving natural resources. Boulder was one of the first twenty communities in the United States to offer curbside recycling.[24] Eco-Cycle is a nonprofit social enterprise with seventy employees and annual financial throughput of $5 million. Eco-Cycle has five divisions:

- the operations of the new county-owned Boulder County Recycling Center;
- the provision of business recycling/composting and hard-to-recycle collection services to more than eight hundred Boulder County businesses;
- the constant creation of new recycling opportunities, such as the Center for Hard-to-Recycle Materials (CHaRM)—Colorado's first community recycling center for electronic waste and many other hard-to-recycle materials;

- the creation of award-winning educational programs to over 25,000 kindergarten through twelfth grade students in approximately one thousand Boulder Valley and St. Vrain Valley classrooms annually;
- the provision of educational services for the community, including a Boulder County recycling hotline and an army of more than 750 volunteers countywide.[25]

Eco-Cycle describes its mission as "to identify, explore and demonstrate the emerging frontiers of sustainable resource management through the concepts and practices of zero waste. We believe in individual and community action to transform society's throw-away ethic into environmentally-responsible stewardship."[26] The current director of Eco-Cycle is Eric Lombardi, who provides his economic philosophy of this operation in chapter 16. Lombardi told me more about CHaRM, of which he is rightly proud. It handles such nontraditional recyclables as block styrofoam, large durable plastics (like toys), TVs, textiles, yoga mats, and bicycle tire tubes. So many items, in fact, that this facility claims to collect more specific items for recycling than any other site in the United States.

Austin, Texas

As I was finalizing this chapter I received an e-mail from a colleague containing information about new programs underway in Austin, Texas. This city has announced it wants to decrease the amount of trash it sends to landfills by 90 percent by 2030. According to the *Austin Business Journal*, the city plans to "establish reuse centers and drop-off facilities, improve its single-stream recycling program by accepting a wider variety of materials, conduct a pilot curbside program to pick up yard waste, food scraps and compostable paper; develop a new Household Hazardous Waste Facility and institute universal recycling and composting requirements to all waste generators in the city." The city was changing its waste management system and its mindset: moving from waste collection to resource management, and in the process crafting a roadmap to zero waste.[27]

Italy

Grassroots activists and local decision makers in Italy have made more strides than in any other nation to bring zero waste strategies to their cities and towns. In fact, so many zero waste initiatives are afoot in Italy that some in the global movement refer to zero waste as the "Italian method."

In more than fifty visits to work on this issue in Italy, I have watched remarkably creative zero waste advocates in action (see the sidebar for a conversation between two of them). I have also seen a variety of different regional approaches. In short, one can find in Italy excellent zero waste models for a host of different-sized community situations, from urban to rural.

Capannori: The Cradle of the Zero Waste Movement in Italy

The Capannori program consists of door-to-door collection of discarded materials on a six-day-a-week basis. Each day the same trucks pick up different designated material. Different regions of the town have different pickup schedules. For example, in one region of town, on Monday the trucks pick up the separated organic materials (kitchen scraps); on Tuesday, paper; on Wednesday, the residual fraction; on Thursday, bottles and cans; and so on. In addition to door-to-door collection the program has a reuse center where citizens can drop off furniture and other reusable items and a zero waste research center where the residual fraction is analyzed and studied. Recently, a pay-as-you-throw (PAYT) system has been introduced to charge for the residual fraction; the more waste people make, the more they pay.

On many levels, it is a model for others around the world to emulate, and it has been a trendsetter from the earliest days of its zero waste campaign. In fact, the history of the zero waste movement in Italy is in large part the history of the zero waste campaign in Capannori. It was the first Italian town to declare zero waste, the first to set up a zero waste research center to analyze the residual fraction, and the first to support a community reuse center.

People often ask why it is that this town located near Lucca—of which few people outside Tuscany, and certainly outside of Italy, have ever heard—has led the way in this movement. There is no single answer to this. Many individuals and organizations have contributed to this very special piece of history. It has been a remarkable team effort.

First of all, there is Rossano Ercolini and the group of his friends and colleagues who make up Ambiente e Futuro. This group was first formed to fight the incinerator proposed for the area. After beating it, Ambiente e Futuro went on to help other communities fight off incinerators. In the process they helped to define and pioneer an alternative solution to their waste problem. Equipped with a one-room office with no heating, no toilet, and few resources, this little group has helped to organize and educate the local community and eventually the whole of Italy on zero waste (see figure 7.1).

Second, Capannori Mayor Giorgio del Ghingaro has had the courage to support zero waste initiatives every step of the way, even though in the early days other politicians ridiculed the concept as "utopian." Now he has become quite a famous person in Italy. Most recently, Ghingaro encouraged many of the other Italian zero waste mayors (now over one hundred) to form an Italian association of mayors for zero waste. Such networking has helped Capannori remain the hub around which the zero waste movement revolves in Italy.

The third key member of the Capannori team is the local waste company, ASCIT, which genuinely believes in what Ercolini and Ghingaro are trying to accomplish. This company has solved every challenge that has been presented to them, the latest of which is the introduction of a PAYT system for the residual fraction.

FIGURE 7.1. The group Ambiente e Futuro was present at a big rally protesting incineration in Florence. The English translation of the banner reads, "DON'T BURN OUR FUTURE." *Photograph courtesy of Pier Felice Ferri*

A Conversation Between Two Activists

I first met Rossano Ercolini in Lucca, Tuscany, on my first visit to Italy in 1996. Ercolini, a primary school teacher, and several colleagues had set up a small group called Ambiente e Futuro (Environment and Future) and were fighting a local incinerator proposal. For about twenty years this remarkable man, without payment and with little recognition (at least at the beginning), first led the anti-incineration campaign throughout Italy and now champions Italy's exciting zero waste movement. Today, he is sought after by communities all over Europe and beyond who want to hear more about zero waste. He promotes zero waste nearly every day—and yet he also carries on with his first love, teaching at the local primary school. With the strong support of Mayor Giorgio del Ghingaro, he has made sure that his hometown of Capannori has taken the lead in many of the initiatives on zero waste in Italy. A superb organizer, it is no surprise that Ercolini was the European winner of the 2013 Goldman Prize (see figure 7.2).

Patrizia Lo Sciuto, a professional dancer, splits her life between Trapani in Sicily and Paris. I remember my first visit to Trapani, a seaside town located about an hour from Palermo, because it was so unusual. Hitherto in Italy nearly every town I had spoken in was at the invitation of a group fighting an incin-

FIGURE 7.2. Rossano Ercolini giving his acceptance speech in the San Francisco Opera House on March 25, 2013, after receiving the Goldman Prize. This prestigious prize was awarded to him for his work battling incinerators and promoting zero waste throughout Italy. *Photograph courtesy of Goldman Foundation*

erator proposal. But Trapani was different. At the time it had no incinerator proposal, and the organizers didn't want to hear the arguments against incineration; they wanted to hear about zero waste so they could push their community in a sustainable direction. Since that first meeting Lo Sciuto has been one of the stars of the campaign to bring zero waste to Sicily. At this point in time she has helped to get eight Sicilian mayors to adopt this strategy. More are on the horizon. She has also helped hugely in the campaign to get zero waste adopted throughout Italy as co-coordinator of Zero Waste Italy.

Patrizia Lo Sciuto (PLS): When and how did this zero waste adventure all start?

Rossano Ercolini (RE): It started for me in 1994 when the Tuscany Region gave a commissary full power to construct an incineration plant in the Lucca plain and the designated spot was Capannori, my hometown. The population immediately mobilized, and environmentalists and committees of citizens launched a campaign. Unfortunately, at that time, almost all the major environmental groups (except for Greenpeace and the World Wildlife Fund) supported incineration. Thus, the mobilization appeared isolated and lacking in scientific and technical support. Without this support it was too easy for the incinerator proponents to dismiss the movement as simply reflecting the "not in my backyard" (NIMBY) syndrome.

At this time, apart from a group called Democratic Medicine, the mainstream medical world in Italy did not adopt a critical stance toward incineration and accepted the promoters' claims that a "state of the art" incinerator posed no health threats. For these reasons we decided to look for international experts to support our gut feeling that even a modern waste incinerator posed unacceptable health risks and that there were viable alternatives to this disposal technology.

Barry Commoner suggested a professor of chemistry at St. Lawrence University in Canton, in northern New York State. This turned out to be Paul Connett. We met him for the first time in January 1996 in Pietrasanta, where local activists had organized a small meeting. But the keynote event was the following morning at the Lucca provincial palace, where a large group of citizens and administrators listened in shocked silence to the first scientific presentation given in Italy by a highly qualified chemist who was opposed to incineration.

PLS: Since those early days Paul Connett has come to Italy over fifty times, giving lectures in every corner of the country. Do you feel there have been real changes in the country as a result?

RE: It is very difficult to perceive the processes of *real* change, but I would say that both the negative aspects of waste combustion and the fact that alternatives are available have spread across the country. The awareness that "no risk is acceptable if it is avoidable" has also grown, as has the belief that communities can defend their democracy not only with words but also with the concrete actions they can take with the waste they produce every day. In a very real sense zero waste is allowing citizens to take back their towns.

PLS: Do you think that by adopting a zero waste strategy administrators may have a tool that creates an effective bridge between citizens and institutions?

RE: Yes, definitely. Things have changed dramatically from the days when citizens who were opposed to incineration were portrayed as the irresponsible enemies of progress. At that time waste management was automatically considered a "technological problem" and therefore solvable only by technicians and administrators. Now it is clear that even administrators are recognizing that citizens are key players in solving the waste problem locally as well as being key partners in the more sustainable changes involved in zero waste.

The same phenomenon, but in much more acute terms, has been observed in recent years in the Naples area, where with respect to waste management citizens have been treated as the enemy. For example, a very unpopular incinerator built in Acerra had to be opened with the military standing by. In addition, a military camp has been installed in a landfill operating in the national forest near Vesuvius. Solutions have been forced upon citizens in a very undemocratic manner using emergency laws and government police powers. But I am happy to say that the situation has dramatically changed since October 2011, when Naples declared a zero waste strategy.

PLS: Rossano, during these years since the mid-1990s a real movement has emerged made up of ordinary citizens, activists, and also resourceful administrators. It stretches out across the whole of Italy. How did you manage to get the attention of so many universities, students, engineers, doctors, scientists, citizens, and administrators?

RE: There have been some insights and strategies, which have been well understood and well aimed. The first was to connect the local struggles spread out across Italy to the international movements against incineration and toward zero waste. Before the 2009 Zero Waste International Alliance conference in Naples [which brought together experts, administrators, and activists who came from around Italy and also from the Philippines, Thailand, India, the United States, Belgium, the United Kingdom, Bulgaria, and Spain] we had brought other international experts to Italy. These included Rick Anthony [see chapter 23], Eric Lombardi [see chapter 16], and Jeffrey Morris [see chapter 21]. Working with these organizations and international experts gave greater credibility to our initiatives.

The second was to define the proper role of experts: they shouldn't replace communities and citizens, but rather—when willing—be there to serve their needs. I quote one of the metaphors that Paul often uses to explain the true role of experts. He says effecting change is like driving a nail through a piece of wood. The expert can sharpen the nail, but you need the hammer of public opinion to drive it home. Over the last few years we have had a growing number of experts help sharpen the nail, and with the help of countless local activists we have produced quite a hammer of informed public opinion to drive the message of zero waste home throughout Italy.

PLS: The capitalist-consumerist system that dominates the globe represents a force that at times seems unbeatable. How can we as citizens, who advocate such simple actions as separating waste, not appear like Lilliputians compared to industrial giants and trade monopolies that offer us their "magic" machines, which have the support of huge government subsidies?

RE: The zero waste path combines a large vision of sustainability with local practical actions. We start with simple steps, which involve everyone. First, each citizen has to avoid the commingling of discarded materials, which makes "waste" (i.e., source separation). Second, these separated materials must be collected with well-organized door-to-door collection systems. These two steps act like capillaries feeding the separated materials into suitable plants for recycling and composting. In this way, a simple democratic starting point, which is so important for the planet, can simultaneously create jobs and stimulate the local economy. The other crucial step must be to involve production managers. It is with them that the industrial and distribution roles in the consumerist chain can be intercepted without too much controversy. The message that has to be addressed is that, while responsible and mature communities do their best to maximize resource conservation and minimize waste disposal, manufacturers and retailers must also play their part. As Paul has said many times, communities have to tell industry that, "If we can't reuse it, recycle it, or compost, industry shouldn't be making it."

PLS: To many zero waste seems a utopian vision. Do you think it is attainable?

RE: You could also call the need to move toward a sustainable society as utopian, but we have a moral obligation to do it. Many people a few years ago would not have believed today in Italy that we would have over two hundred communities diverting more than 70 percent from landfills and some small communities reaching over 80 percent. I am not saying that it is going to be easy to divert the remaining 20–30 percent, but like sustainability there is a moral imperative that we try. At the same time we have to denounce the unsustainable economic and social model of the throwaway society, which is propped up by waste disposal options such as landfills and incinerators.

The zero waste path combines a utopian vision with the concreteness of simple but practical actions. The rediscovery of being part of a community, which includes citizens, experts, business owners, and administrators, is a thing of beauty. By the simple process of doing what is necessary to put material things back to their rightful place the community recognizes itself again. It's not a coincidence that in the United States one of the most significant sayings of the grassroots is "A threatened community is a strengthened community." Especially if we all work together to fight the threat.

Zero waste represents a *fight against* toxicity and a *fight for* sustainability. We first see the toxicity in local forms of pollution such as ash, dioxins, and toxic nanoparticles released by incinerators. But then we realize the larger threat these incinerators pose to the global community by hiding our lack of sustainability from general view.

Eric Lombardi when he was touring Italy said that the wars of the future are likely to be wars over resources, and thus zero waste is the new peace movement. A war against waste is an urgent war against wasting resources, and thereby a way of avoiding a far more tangible and dreadful war in the future.

It's not an exaggeration to say that zero waste can be summarized as a message of love: a love for each other, a love for our communities, and yes—even a love for our planet.

PLS: In Sicily citizens have been able to stop the building of four proposed incinerators for the island, and as of 2012 eight communities have declared zero waste. One of those incinerators was planned for the lush farm area of the Paternò valley near Catania. It was also located in an area of great natural beauty and of great archeological interest. The contrast between the beauty of the Paternò valley and the ugliness of incineration is extreme. That valley nourishes life with wholesome food and has been doing so for over two thousand years. Burning waste on the other hand is a form of capital punishment for finite resources. Some men would reduce these resources to ash, but I believe that discarded materials are entitled to a new existence. They need new life, not death by fire.

RE: In the Paternò valley, citizens organized themselves into committees to oppose the proposed incinerator. I must pay tribute to Paolo Guarnaccia, an organic farmer from the Vivisimeto committee. He was a key leader in the battle. Paolo has gone on to become one of the leading players in both the anti-incinerator and zero waste movements in Italy.

Along the way, ASCIT has been able to adjust their program to take into account both economics and convenience to the citizen, as seen in a recent move to collect glass separately from plastic and cans. With this separation each material yields a higher resale price (180 euros per ton for plastic and 20 for glass). Before this development ASCIT had to pay 49 euros per ton to dispose of these mixed materials because they did not meet national quality standards. To achieve this change, ASCIT has elegantly rescheduled the calendar for pickup. It now picks up the glass once every fifteen days and the plastic and cans three times every two weeks. This move has saved the community 400,000 euros per year.

From the very beginning ASCIT has excelled in communicating to the public what they needed to do to make the program work. Their communica-

PLS: Your small group, Ambiente e Futuro, has opened up a window onto a reality previously unknown in our country. What helped to open the windows to this historical moment?

RE: You have to start locally. If there hadn't been a local struggle against incineration there wouldn't have been this opening. At the same time our approach has always aimed to address local issues without giving into "localism." Without a global consciousness, your struggles live or die alone. Reaching out to others has been the key. Reaching out to other communities, reaching out to experts outside our community, reaching out to administrators and entrepreneurs with vision and joining global networks like GAIA and ZWIA have all been critical.

PLS: Is it possible that these local battles and global zero waste movement can lead us beyond the perverse system of consumerism?

RE: I think the zero waste movement has a huge role to play on the global stage. More than anything else what we are asking for is for political leaders and other institutions to work with the grassroots to secure communities that improve the present by consciously sharing things with each other as well as sharing with the future. The message is fairly simple: When you throw something away or, worse, when you are making something to throw away, you are throwing the future away. Most people get this message when they realize that what we are talking about here is the future of our own children and grandchildren.

Ultimately, we all have to have a faith that what people really need to be happy are not throwaway objects but permanent relationships with other people in their community. To make this global we need to produce successful local models, which others can copy. These models need to show that we can avoid wasteful disposable patterns while pursuing the larger goals of social justice and environmental sustainability and be happy doing it.

tions (see figure 7.3)—whether they be posters, leaflets, or booklets—have always been very clear, colorful, creative, and humorous.

The fourth member of the Capannori team is Alessio Ciacci, who is employed by the council to oversee the program and act as outreach to administrators in other communities that want to copy the Capannori model. In 2013, Ciacci won a prestigious environmental award for his zero waste promotional efforts.

Finally, the program has received strong support from the local business community and local institutions in terms of various waste reduction initiatives. Local farmers have provided dispensing systems so that citizens can refill their own milk bottles from refrigerated units (see figure 7.4). A local company (Ecobimbi) has designed and produced reusable diapers for babies and adults,

	lunedi	martedi	mercoledi	giovedi	venerdi	sabato
PRIMA SETTIMANA						
Rifiuto non riciclabile						
Organico						
Carta e Cartone						
Multimateriale						
Vetro						
SECONDA SETTIMANA						
Rifiuto non riciclabile						
Organico						
Carta e Cartone						
Multimateriale						
Vetro						

FIGURE 7.3. Inviting graphics and simple instructions are keys to success, as in these ASCIT publications. *Images courtesy of ASCIT*

and local pharmacies have provided the first reusable diaper (*pannolini*) to new parents free of charge. A new food store (Effecorta) provides food and wine from the local area using dispensing systems for every product, thus eliminating the need for plastic bags and bottles. In all it sports sixty taps for all kinds of liquids (from detergents, shampoos, milk, and water to honey, olive oil, and wine) and sixty dispensing units for beans, pasta, and other solids (see figure 7.5). In addition, local schools are replacing disposable plastic tableware with china, glass, and stainless steel.

The Capannori zero waste research center has investigated the nonrecyclable items in the residual fraction. These have led to the local production of reusable diapers instead of disposable ones, a more durable shoe design, and negotiations with the giant coffee company Lavazza to seek an alternative to their disposable one-serving coffee capsules. The research center is also promoting the reuse of the plastic fraction remaining after the recycling of paper in local paper mills (Capannori is the capital of the European paper industry). It is hoped that local industries can use the recovered plastic along with some short cellulose fibers to make plastic boards for use in construction. The center has also made a "little museum of badly designed packaging" that they display at meetings.

According to Ercolini, "Lucca province [where Capannori is located] has become the cradle for sustainable waste management in Italy," taking over the mantle that Capannori pioneered. It is the only Italian province that has stopped all incineration and has announced that it does not want any more incinerators built. Meanwhile, the closed refuse-derived fuel (RDF) burning

FIGURE 7.4. Local dairy farmers provide refrigerated milk-dispensing systems in Capannori so that residents can use and reuse their own milk bottles. *Photograph courtesy of Pier Felice Ferri*

FIGURE 7.5. Some of the sixty taps available to deliver a variety of liquids to customers using their own containers in Effecorta. Customers can sample eight different wines before choosing to fill their own bottle. *Photograph courtesy of Pier Felice Ferri*

facility in Pietrasanta is going to be converted into a residual waste screening facility. In addition, eight municipalities in Lucca province have adopted a zero waste strategy—and in 2012 the Lucca municipality itself officially adopted a zero waste goal.

Sicily: The Fight Against Incineration and for Zero Waste

As we have seen in other countries it is frequently the passion that is engendered fighting incineration that feeds into the grassroots efforts that push for zero waste. This was particularly true of Sicily.

Sicily is an island that has fascinated many people throughout history. It is rich in contrasts, both in its natural beauty and in its culture. Over the centuries the island has inspired many artists, painters, poets, philosophers, filmmakers, and musicians. Sicily is a land of rulers and ruled, of conquerors and conquered. Since ancient times the conquerors have sought to impose their language, culture, customs, and traditions on the conquered. The most recent imposition was led by the incineration industry that sought by stealth to impose its technology on the island—specifically, to construct four incinerators to burn all the island's discarded resources. The deal was worth five billion euros to the incineration industry. Magistrates who have since investigated this plan have discovered a huge tangled web of self-interest involving strong cooperation between politicians, the Mafia, consultants, and entrepreneurs. This web of intrigue stretched from Sicily throughout Italy and beyond.

It all started in 2002 with a call for tenders for the construction of incinerators that were to be located in Palermo, Casteltermini, Paternò, and Augusta. This scheme was organized by Salvatore Cuffaro, who was the president of the Sicily Region and the commissioner appointed for handling the waste emergency. Cuffaro was arrested in 2011 for aiding and abetting the Mafia.

For the purposes of waste management, President Cuffaro divided Sicily into four regions, like a big cake. Each region was required to set up a joint enterprise (code words for an incinerator) to manage all of its waste. The combined capacity of the four facilities was more than the total amount of waste produced by the whole of Sicily (2,604,410 versus 2,576,660 tons per year!). This was a clear violation of European directives that required following the waste hierarchy of reduction, reuse, and recycling before waste disposal options.

The contracts to build these incinerators were won by four energy conglomerates: Pea Platani, Energia Ambiente, Tifeo, and Sicily Power. Building was started in 2007 but was quickly blocked by a ruling of the European Court of Justice (ECJ), which canceled the contract for the failure to comply with the public tender regulations set by European directives.

The choice of where to build these plants was rife with problems. An incineration plant near Augusta was to be located in one of the most polluted industrial areas in Europe. Attempts to site incinerators in already polluted areas are not uncommon—a practice that some refer to as the creation of sacrifice areas. The cancer rate from environmental pollution in Augusta was already very high, and the idea of adding to this via a huge incinerator led to massive opposition from the local population. I remember speaking to a huge audience in Augusta on this subject, and the intensity of feelings on the matter was palpable. "*Basta, basta*" ("Enough is enough") was the unified cry that echoed around the hall.

The other three incinerators were to be located in areas that violated local planning laws designed to protect areas of high natural beauty. Indeed, it is difficult to find a place in Sicily where building an incinerator would not violate these local laws. But that didn't get in the way of the Cuffaro government.

During the first few years of the attempt to implement the Sicilian president's plan each incinerator proposal triggered the formation of a citizens' committee dedicated to fighting the project. Despite the citizens' willingness to entertain an alternative strategy involving the building of centers for recycling and composting, President Cuffaro made no genuine effort to move in this direction. Press reports revealed that he had a personal economic interest in the incinerators.

After the resignation and arrest of Cuffaro, a new president—Raffaele Lombardo—was elected. In the beginning, he continued to support the idea of burning the waste. Meanwhile, the local committees of citizens who were protesting against the construction of the four incinerators were joined by other nongovernmental organizations (NGOs) promoting zero waste.

In April 2009, President Lombardo was to discuss the announcement of the four new incinerators to the Sicilian parliament. However, on the day he was to do so, zero waste activists and many other citizens' groups and associations gathered under the windows of the meeting hall of the Palazzo d'Orleans in Palermo. Over megaphones, citizens pleaded with President Lombardo to use his common sense and reject the incinerator plan. Meanwhile, with the help of GAIA thousands of e-mails were sent by activists from all over the world asking Lombardo to cancel the incinerator plan and instead implement a zero waste program. GAIA supporters argued that Sicily had great potential for the application of practices that were not only better for the local and global environments but also had the potential for creating many jobs on the island.

Months later, much to the joy of zero waste activists but to the dismay of the powerful forces behind the incineration plan, the government announced the cancellation of the incinerator contract. Everything changed from that day onward. President Lombardo declared to the press that "incineration is a bankrupt technology."

In July 2012, after several years of discussions, a new plan was approved by the Sicilian environmental ministry and the Italian government. The goal of the plan is to achieve 65 percent waste diversion for the whole island, including the introduction of source separation and recovery of organic waste—a significant signal that Sicily plans for sound management for the bulk of its waste. It is expected that the remaining part will be handled in mechanical and biological treatment facilities. However, there is also the possibility that some of the waste will be burned as refuse-derived fuel (RDF) in cement kilns. But this latter determination will be up to local authorities. Meanwhile, citizens are pushing more and more communities to announce zero waste plans that would avoid any form of burning.

Eight Communities Show the Way

In the last three years while the island government was inching ever so slowly toward recycling and composting, a number of municipalities have taken the lead in aggressively pursuing a zero waste goal. These have pioneered a number of best practices. In all, eight mayors have adopted a zero waste strategy for their municipalities.

MARINEO (PALERMO)

The municipality of Marineo (population 6,814), located in the Palermo region, embarked on a zero waste strategy after having made some important steps in response to a failed system of waste management.

- In September 2009, the Center for Composting of Castelbuono (Palermo) was opened and the door-to-door collection of plastic, glass, metals, paper, and organic waste was started (this excluded the residual fraction that was still collected in drop-off containers). Volunteers distribute and explain the use of the brown containers designed to receive kitchen scraps for composting.
- In March 2010, the residual fraction was included in the door-to-door collection system. This meant that all the drop-off containers were eliminated from the streets.
- In February 2011, Marineo formally adopted a zero waste program.
- In April 2011, Marineo reached 52 percent diversion and was able to decrease local taxes by 16 percent.

In addition to door-to-door collection the city of Marineo has adopted several waste reduction initiatives. These include the promotion of reusable diapers and reusable shopping bags.

The municipality has demonstrated that the new program is saving the taxpayers money. In 2008, the municipality of Marineo spent 302,000 euros for disposal of their waste in the landfill at Bellolampo (Palermo). However, in 2011 it spent only 110,000 euros for the door-to-door collection and the recycling of recovered materials.

I have visited Marineo three times, and each time the excitement has grown. Mayor Francesco Ribaudo has managed to involve the whole community. On my last visit a local composer had written a zero waste song and distributed free CDs to the crowd; after the formal presentations we were treated to a delightful concert of music from young musicians, and after this everyone was treated to food and wine produced by the local community. The mayor believes that zero waste is important for both the local culture and the local environment.

COLLESANO (PALERMO)

Collesano (population 4,254) is located in the Madonie mountain range, a large territory with both a dense population center and a highly dispersed rural area. This combination makes it difficult to design a single collection service applicable to all homes. The municipality has chosen a mixed system: the center has organized door-to-door collection of wet and dry waste, and for the rural area it has built recycling centers (ecological islands) where citizens can drop off their dry separated waste—paper, plastic, glass, and metal. For the organic portion the municipality has provided garden composters to any citizen who has requested one.

Several additional projects are planned for Collesano.

- A sustainability park is planned for a central green area. There citizens will be able to barter their unwanted reusable objects with the city for local food products or other basic necessities. The goal is to make the practice of reuse and recycling both economically attractive and convenient for citizens.
- A composting plant will be built in the rural area to manage the organic wastes of Collesano and other neighboring municipalities.
- The municipality of Collesano has managed to stop the building of a pyrolysis system (pyrolysis is a process where an external heat source is used to convert solid waste into a gas, which is then burned to produce electricity) that was planned for the enchanting Madonie Park. The mayor and his environmental consultant appointed a technical committee with local administrators. This committee invited Enzo Favoino, one

of the leading experts on waste management in Italy and Europe, to help with designing their program. Favoino has shown, with a detailed analytical assessment, that a "cold" mechanical and biological treatment system is far superior to a pyrolysis plant for recovering material from the residual fraction in addition to the material recovered in the curbside collection and drop-off systems.

CALATAFIMI-SEGESTA (TRAPANI)

Calatafimi-Segesta (population 7,258), located in the Trapani region, is yet another small Sicilian town located in a picture-book landscape (see figure 7.6).

Calatafimi-Segesta's zero waste program began in 2010. Since then the city has:

- passed local ordinances requiring source separation and door-to-door collection;
- designed educational programs to motivate children in elementary and middle schools, business owners, and the general citizenry;
- organized public meetings and conferences to gather input from citizens and communicate the benefits and progress of the program in terms of reducing the waste tax;
- built a composting facility on land confiscated from the Mafia (this plant, the biggest of its kind in southern Italy, has a capacity to handle 36,000 tons of organic waste and serves eleven municipalities in the Trapani province);
- built a drop-off municipal collection center for the recyclable fraction;
- signed a memorandum of understanding with the waste collection companies for the removal of asbestos (economically attractive for citizens);
- purchased a public water fountain to reduce the number of plastic water bottles used by citizens;
- provided economic support for the use of compostable tableware (made out of Mater-Bi) for fairs and festivals;
- purchased a bio-shredder for the preparation of mulch and compost to enhance the fertility of the local soil;
- copied Capannorri in attempting an analysis of the residual fraction.

ALCAMO (TRAPANI)

The municipality of Alcamo (population 45,985), also located in the Trapani region, has taken important steps to raise awareness and motivate citizens and traders to support recycling. Currently, they have achieved a diversion rate of 50 percent. They've also stimulated and supported regional activism by hosting

FIGURE 7.6. The municipality of Calatafimi-Segesta. *Photograph courtesy of Patrizia Lo Sciuto*

FIGURE 7.7. Some of the more than one hundred attendees of the zero waste conference held in Alcamo on March 24, 2012. This picture includes all eight mayors of the Sicilian cities that have declared zero waste. *From left to right standing:* Francesco Gruppuso, Nicolò Ferrara, Luca Gervasi, Massimo Fundarò, Piero Cannistraci, Patrizia Lo Sciuto, Paul Connett, Aimée Carmoz, Giuseppe Glorioso, Giacomo Scala, Vincenzo Emanuele, Francesco Ribaudo, Enzo Favoino, Giusy Cicero. *From left to right sitting:* Rossano Ercolini, Pietro Daidone, Paolo Guarnaccia, Concetta Mattia, Silvia Coscienza. *Photograph courtesy of Pier Felici Ferri*

an important meeting on zero waste for the whole of Sicily—gathering activists, experts, and all mayors whose towns have declared zero waste goals (see figure 7.7). I was very pleased to attend this conference and meet in person all eight mayors who have pioneered zero waste in Sicily. Some of these mayors are every bit as enthusiastic about the promotion of their programs as any grassroots activist. One of the pleasant aspects of this daylong conference was the luncheon in which each mayor brought to the table wine, cheese, meat, and other local food products from their towns. The whole day was a delightful mixture of work and fun.

BUSETO PALIZZOLO (TRAPANI)

Buseto Palizzolo (population 3,095) is another rural town in Trapani (see figure 7.8). Its program has included a pilot project on home composting, which was conducted because of the expense of organizing door-to-door collection of organic waste. Backyard composters were distributed to households with a garden. The municipality has also developed a project called Viva Paper for all schoolchildren in the town. This involves a competition to see who can collect the most paper, cardboard, and old books for reuse and recycling. The municipality is also building a plant for the treatment and separation of dry recyclables for the whole territory.

BIANCAVILLA (CATANIA)

The municipality of Biancavilla (population 23,947) in the province of Catania is located in the foothills of Mount Etna. In less than one year, the city has

FIGURE 7.8. The municipality of Buseto Palizzolo. *Photograph courtesy of Patrizia Lo Sciuto*

dramatically changed the waste collection system. The rates of recycling have increased from 6 percent in August 2011 to 64 percent in July 2012. Grassroots activists in the Zero Waste Biancavilla Association have contributed to this important change by generating a greater awareness among both citizens and administrators about the possibilities of the program. This has included volunteers going door-to-door to help motivate citizens to achieve the best results possible.

GRATTERI (PALERMO)

The municipality of Gratteri (population 1,081) is located in the Palermo region. Their program is still in its earliest days and has only reached a modest diversion rate of 36 percent. However, it has created a significant awareness campaign on recycling and waste reduction and hopes soon to catch up with some of the diversion rates achieved by the other zero waste towns.

CASTELBUONO (PALERMO)

I have saved the most interesting of the eight zero waste communities in Sicily for last. Castelbuono (population 9,301) is located in the Palermo region (see figure 7.9). With steep hills and narrow streets it needed a creative approach to solve the problem of door-to-door collection. Former mayor Mario Cicero, a man full of enthusiasm and character, found the answer—donkeys.

Mayor Cicero linked the recovery of a traditional donkey breed from Sicily—called Ragusa—with the challenges of waste collection in the narrow streets and steep hills of the town. Today, the donkeys with their handlers go door-to-door to collect separated materials.

Joan Marc Simon, the European coordinator for GAIA, visited Castelbuono in March 2012. He expressed a positive opinion of the program to members of Zero Waste Europe, writing that "by making the donkeys useful, recovering this breed stops being a cost and becomes an asset." He adds that the economics are excellent. According to Mayor Cicero, a small truck to collect waste costs more than 15,000 euros and lasts five to ten years—whereas a donkey costs less than one-tenth that amount and lasts longer. The neighboring municipality of Cefalú made a different choice. It chose modern waste collection methods—trucks that pick up roadside containers, a choice that (according to Simon) has generated millions of euros in debts. Castelbuono, on the other hand, has a balanced budget—a rarity in the area.

Simon points out two more valuable uses of these donkeys: their milk is highly valued, being the closest thing to human milk that is available commer-

FIGURE 7.9. The town of Castelbuono. *Photograph courtesy of Joan Marc Simon*

cially in the area. And the donkeys are used for donkey-assisted therapy, a powerful tool for treating mentally disabled people. According to Mayor Cicero, "The Ragusa are once again participating in ceremonies and other local festivities without extra cost to the community." He explains that elsewhere in Sicily the government has spent 1.5 million euros to reintroduce the Ragusa donkey but is not doing anything with them—they are a net cost. Instead, Castelbuono started with four and now has forty-five donkeys, which pay for themselves. Residents in Castelbuono know the donkeys by name and often stop to feed them when they pass by (see figure 7.10).

During his visit to the town, Simon accompanied the handler who worked with a donkey named Valentina. Before the donkey program, the handler was

FIGURE 7.10. The donkeys in Castelbuono receive greetings and treats from local residents and tourists alike. *Photograph courtesy of Joan Marc Simon*

FIGURE 7.11. Some donkey handlers have found their lives turned around by working with the donkeys and conversing with tourists. *Photograph courtesy of Joan Marc Simon*

having problems with depression and addiction and would frequently not show up for work for five days in a row. Since he started working with Valentina, he has not missed a single day of work and has managed to reorganize his life. The bond with Valentina and the daily contact with neighbors and tourists have managed to bring him back into society (see figure 7.11).[1]

Personal stories like this are just the tip of the iceberg of a successful story of integration, education, creativity, and sustainability in Castelbuono. Separate waste collection and diversion rates in Castelbuono are higher than in any of the neighboring towns.[2]

Castelbuono hosts a composting plant that treats the separately collected organics of the province, and it is a proud member of the Italian network of zero waste municipalities. Many other zero waste municipalities with similar characteristics inside and outside Sicily are studying this successful and important model for the zero waste strategy.[3]

Naples

For nearly ten years Naples epitomized the very worst waste management system in Italy. No effort was made to source separate, and garbage piled up on the side of streets. Periodically, angry citizens set fire to the waste, producing gruesome images captured by TV stations around the world. Meanwhile, aided by the central government and ignoring pleas from grassroots activists, the Naples authorities continued to push to build incinerators and more landfills in the region. This led to huge opposition from citizens. Not only did they not trust the authorities to run the incinerators properly, but they were also very wary of any landfills near their homes because it was well known that the Mafia had been using these for getting rid of hazardous waste from all over Italy. On top of all this, scientists were finding very high dioxin levels in local milk supplies as well as finding very high cancer rates in the region.

One other "solution" adopted by the Naples government was to pack shredded garbage into huge plastic bags called *ecoballe*. These were then piled up in huge pyramids in local fruit orchards. I once commented at a press conference that you "couldn't grow peaches on *ecoballe* but you could on compost" and argued that the crucial step in solving Naples's trash problem was to organize source separation of clean organics for composting. However, in about ten visits to Naples, while I got to speak to the media and at many meetings organized by citizens, I never got to speak to the decision makers in the Naples government. Thus it came as a huge and delightful surprise when I was invited to go to the municipal building on October 3, 2011, and present the zero waste strategy in front of the newly elected mayor and vice mayor. It was at this meeting that Naples declared it was going to opt for a zero waste program.

It was a giddy moment for all of us. Two additional pieces of information help to explain this extraordinary development. First, the mayor, a former senator in the Italian parliament, had already established that he was willing to stand up to the Mafia. Second, the vice mayor had been involved in a local zero waste campaign before being elected.

Though it first took root in a small corner of the nation, one

FIGURE 7.12. Multiple voices promote the zero waste message in Italy. The Hotel Association of Capri, for instance, produced this comic book—in which author Paul Connett is portrayed as Aladdin's genie! *Illustration courtesy of Hotel Association of Capri*

A Mayor's Perspective

BY TOMMASO SODANO, VICE MAYOR OF NAPLES

The following has been excerpted from the preface written by Tommaso Sodano for the book (published in Italian) *Zero Waste: A Revolution in Progress*.[4]

In the first few months of our new administration we threw out the old emergency management of waste (involving siting landfills and incinerators in the Naples area) and set up plans for the future based on environmental protection, resource conservation and sustainability. One of our goals is to extend door-to-door collection throughout the city. The first step in this effort was to move from 140,000 being served in this way to 325,000. We will try to bring this up to 600,000 residents by the end of 2012.

The framework outlined by our zero waste resolution, established in 2011, is much larger in fact than door-to-door collection, though. It identifies not only recycling but investment into waste prevention and reduction measures. These include prohibiting the sale and use by commercial operators of disposable containers and plates. We have laid down rules for the collection of commercial waste with penalties for those breaking the rules. We will also be investing in plants to maximize the recovery of materials to avoid building an incinerator in Naples.

To reach out to citizens and increase recycling we have established ecological collection centers in fourteen sites located in the town. We have also had the help of various nongovernmental organizations. For example, Legambiente and other organizations have organized various events. Over four days there were over one hundred initiatives in the streets aimed at informing citizens in the city

can now see the zero waste message being spread far and wide in Italy. In Capri, the hotel association even produces a comic book to get the message out (see figure 7.12). But even though the zero waste challenge has been firmly undertaken in Italy and elsewhere, with goals well beyond just recycling, we would still do well to consider something activist Patrizia Lo Sciuto recalls hearing Mafia boss Vincenzo Virga say in a tapped phone message: "In my recycling plant, waste comes in and gold goes out." Virga, who ran a recycling plant near Trapani, went to prison—but nevertheless understood very well that wastes are resources to recover. Now we are seeing many municipalities in Italy catching up with this same message, but instead of the profits going into organized crime they are staying in their communities and working for the citizens and the environment.

about proper recycling. The common theme of these initiatives was to show that the administration and citizens need to work together to take care of our city. Although we are beginning to see the liberation of Naples from waste, we still have to deal with the structural deficit in the region with respect to facilities to handle two thousand tons of waste per day.

To give us enough time and tranquility to handle our waste in a rational and self-sufficient way—and to avoid another crisis—we are forced to send some of our waste abroad, to an incinerator in the Netherlands. Some see a contradiction with this because of our opposition to waste incineration, but I feel content in saying that this will be a transitional phase, necessary to avoid the presence of garbage in the streets. For eighteen years of the waste emergency we have seen the piling up of our waste on the streets as a way to force landfills or incinerators onto local communities.

We are attempting to identify all the companies that already implement policies to reduce waste and sell products with low environmental impact and to identify all actions that can be applied successfully in the city. These include green purchasing, helping large retailers to install equipment for volume reduction, and establishing a tariff system based on the actual amount of waste produced by households (pay-as-you-throw systems). We are also planning a community center for repair and reuse for packaging and durable goods. In order to make sure this path to zero waste is faithfully followed we have set up an "observatory" to monitor problems and find solutions.

We are very much aware that the experience of zero waste in Naples can be an example for other large cities. Zero waste is not utopian but the only way to go if we wish to be a sustainable city.

Australia and New Zealand

In both these countries we have seen very ambitious zero waste plans falter. However, as with other things in life, we can learn as much from when things go wrong as we can from when they go right.

Australia

Earlier in this book you read about the remarkable spark lit by Australia when, in 1996, Canberra became the first city in the world to enact a zero waste law and news spread throughout the United States and elsewhere, inspiring other efforts. The law required the community produce "No Waste by 2010."[1] The initiative generated tremendous enthusiasm and had the support of public officials.

When I last visited Canberra, in 2004, it was achieving over 70 percent diversion. To be fair, though, this figure was largely influenced by the huge diversion of both yard waste and (very heavy) construction and demolition (C&D) debris. Canberra was undertaking a number of programs at the time.

There was a two-bin door-to-door collection system, with a yellow bin for recyclables and a green one for residuals.

There was an educational effort to encourage backyard composting.

There was a massive commercial yard waste composting operation, which was also receiving other organic waste materials from both agricultural and industrial operations.

There were several large facilities recycling C&D debris (there is a local ordinance that requires those who deconstruct or demolish buildings to identify where the materials are going to go).

There was a very impressive operation selling secondhand building materials obtained from deconstruction enterprises.

There was a paint recycling operation.

There were two large reuse and repair facilities: Aussie Junk and Revolve. The Revolve operation was one of the most impressive of its kind I have visited anywhere in the world, with all the items displayed like a department store—with different rooms for the kitchen, bedroom, and dining room.

There was a site that the government had set up as a nursery for companies that were attempting to use a variety of discarded materials to make new products.

There was also the system in operation at the Mugga Lane Landfill. At the entrance to this landfill citizens and local businesses could drop off separated recyclables, waste oil, and other household toxics and their reusable items at Revolve (Aussie Junk operates at a transfer station at a different location). All this dropping off could be done before householders went over the scales and dropped the remainder of their discards. This was done over a low wall onto a concrete floor, designed to prevent scavenging. However, employees from Revolve had the right to pick out more reusable items from these discarded items. Beyond this scale was the landfill itself.

I was intrigued to see great flocks of ibis, a bird with a long curved beak, scavenging organic scraps from the landfill. In ancient Egypt this bird was considered a sacred creature; it seems to have come down in the world since then. Beyond the landfill was a huge area that was being cleared for what was scheduled to become a resource recovery park. Here the object was to colocate all the businesses that could process the discarded materials and where possible make them into new products. The idea of the resource recovery park was that the government, which owned the land, would provide the infrastructure and rent out space to different entrepreneurs (rather like an airport). The first building, which was nearly complete, was a giant material recovery facility (MRF); the second was to be an education and reception area for visitors. There was already a small prototype of such an education center built close to the drop-off area at the landfill entrance. This small center had been built out of recyclable materials. All of this and more is shown in a videotape I shot at the time.[2] I left Australia in 2004 with very high hopes for the future of this operation.

Sadly, after a change in government, enthusiasm for this program has faltered. In 2009, the new minister for the environment announced a cancellation of the "No Waste by 2010" goal.[3] Some have suggested that the government realized that it would not reach zero waste by 2010 and decided to duck out early. Others have suggested that the government did not want to lose the revenues the landfill earns from those dumping mixed waste, estimated to be about $12 million a year.

There are two lessons I think we can learn from this. Firstly, the Australian Capital Territory (ACT) government, unlike San Francisco, never introduced curbside collection of food waste. Instead, they put their emphasis on encouraging citizens to compost in their own backyards. The problem with this is unless the majority of citizens comply the residual fraction becomes less amenable for further handling because of the odors emanating from putrescible food scraps. With space available it was probably more tempting to expand the landfill than to invest in a residual screening facility as done in Nova Scotia.   Secondly, it underlines the importance of good and committed leadership that is essential to sustain a zero waste program. Ambitious zero waste programs that depend on enthusiastic leadership are always going to be vulnerable to a change in government.

So the current situation is that Canberra has completed the MRF, has expanded the Mugga Lane Landfill, and is even talking about building a facility to convert the residual fraction into RDF. A 2011 citynews.com.au article entitled "Canberra's Waste Dilemma" summarized the situation:

> Following the failure of "Zero Waste 2010", the Legislative Assembly now sits on a philosophical garbage dilemma with paths leading in two directions: One towards the use of waste for energy and the other towards compost.[4]

In this article a spokesperson for the Greens wanted to know why the minister "has not investigated the proven use of organic green bins and their contents to create compost from waste." She adds, "We know Canberrans have accepted their recycling bin quite well, they may well do the organics very well, too. It has worked just over the road in Goulburn, they're doing well with their organic recycling."[5]

So it would appear that the Canberra program may be poised at a crossroads: In one way they might fall victim to the far too cozy relationship between municipal decision makers and the waste industry (described by Helen Spiegelman in her essay in chapter 18) and in the other they may be able to resurrect the original vision with a curbside composting program. Meanwhile, the good news is that the zero waste message set in motion by Canberra has produced more lasting success in other countries. This was particularly true in California and partly true in New Zealand. So Canberra's faltering is less of a loss to the zero waste movement worldwide than it is a surrendering of Canberra's once-worldwide leadership position. That mantle has now shifted to San Francisco.

New Zealand

In the late 1990s Gerry Gillespie, one of the organizers of the Canberra program, traveled to New Zealand to spread the zero waste message there.

A key player in these efforts was Warren Snow, a former businessperson, who worked with the owner of a chain of superstores (The Warehouse) to provide funding to New Zealand communities (via the Zero Waste New Zealand Trust) to help set up zero waste programs.

According to Cecilia Allen of GAIA, by early 2005, 72 percent of the nation's councils had established zero waste targets, and close to 95 percent of the population had access to recycling programs. The councils control waste management through bylaws, legislation, economic drivers such as discounts at transfer stations for diverting materials toward recycling, and contracts that pay for waste minimization—rather than for volumes of waste moved.[6]

However, a word of warning is needed here. Making an announcement that you have adopted zero waste is quite different from taking the practical—and sometimes difficult—strides toward achieving or moving significantly toward that goal. I remember being deeply shocked when I visited Christchurch in 2004 and found that that city, which had formally adopted a zero waste strategy, was actually building a massive regional waste landfill a few kilometers from the city. This was hardly the small interim landfill envisaged as Step 10 in the zero waste program.

The zero waste philosophy was eventually endorsed by the national government when it was enshrined in the Waste Minimization Act, passed into law in 2008.[7] According to Allen, "The Waste Minimization Act made waste levies the law, starting at the rate of $8/tonne for waste going to landfill. Half of the money collected goes to local councils to assist with waste minimization, and after costs are deducted, the rest is to be allocated to recycling schemes."[8]

When Snow became disillusioned with the Warehouse chain's ability to follow through on the sustainability strategies he had helped them develop—including marketing products and packaging that reflected a genuine commitment to social justice and sustainability—he left the firm. The source of money used to stimulate communities to begin and sustain a zero waste strategy dried up, and many communities have now fallen by the wayside. However, there are still a few trailblazers who are setting a good example.

A recent trailblazer is Auckland Council. This new "super" city council created from what were seven adjoining city councils put zero waste by 2040 firmly at the heart of its New Waste Management and Minimization Plan. Furthermore, they have set in place a range of strategies to help them achieve this target—including getting organics out of the domestic waste stream and developing a resource recovery network. This network will involve local community groups, Maori organizations, and local businesses and will ensure that residents and businesses are able to recycle materials and used goods conveniently and economically. The resource recovery network will operate a used goods trading system over the whole city.[9]

Another trailblazer is the Opotiki District Council, which has achieved a remarkable 90 percent diversion.

Allen has found some examples of community reuse and repair centers (Step 5 in the 10 steps to zero waste) in different parts of New Zealand, which she describes as attractive, informative, and even entertaining.

1. The Trash Palace is located in the city of Porirua and is used by teachers, artists, and craftspeople looking for cheap art materials recovered from local industry and businesses.
2. The Recover Store, in Dunedin, is described by manager Kath Abernathy as "not just a place to drop off or buy—it's very much a social gathering place for some people—it's a meeting place."
3. Innovative Waste Kaikoura is located in Kaikoura, on a scenic promontory overlooking the coast. The former manager John Ransley says that one day "people will come to Kaikoura to visit the resource recovery center and then, if they have time, will go whale watching!"
4. A community reuse and repair center in Amberley employs people who would otherwise have difficulty finding productive work, such as youths at risk, people with mental disabilities, and retirees. A local doctor said of this facility: "People with few social contacts come along here—it's a social outlet and sometimes they get a bit of money for their efforts. As a doctor I have actually sent people here. It's not a stretch of the imagination to say that by saving these resources and using previously undervalued people to do it we are improving the health of the community."

Allen also points out that the Zero Waste New Zealand Trust has helped to set up a Zero Waste Academy. This is run by Massey University with the cooperation of the Palmerston North City Council. The academy's activities include developing quality standards for zero waste practice, particularly in compost production, and helping educate local governments on how to develop comprehensive recycling and waste management programs. It also provides opportunities for postgraduate research. Such a facility would be a strong candidate to engage in zero waste research (Step 8 in chapter 2). Allen's findings support what Snow pointed out in correspondence with me:

A strong dose of passion and desire for rapid change still resides within almost all New Zealand communities and for many pioneers zero waste is still the catch cry for change. The initial wave of enthusiasm may have lost its national coherence but it still simmers at the community level and still has its champions in local government and will no doubt steadily consolidate and build on the earlier gains to develop a more sustainable movement. Watch this space, as they say.[10]

The New Zealand experience is a salutary reminder to us that it is not countries but communities that recycle. What we need to look for are model communities (regardless of what the rest of the parent country may or may not be doing) that are providing working examples of one or more of the ten steps to zero waste so that they can inspire other communities to copy them.

Canada

When it comes to waste management, Canada is a really mixed bag. On the one hand provinces like Ontario and British Columbia, which in the past have been associated with progressive environmental policies, while speaking in favor of zero waste have succumbed to building more incinerators. On the other hand provinces like Nova Scotia, which have not been noted for strong environmental policies in the past, have rejected new incinerator proposals and pioneered key steps toward zero waste. In fact, Nova Scotia was the first province in Canada to reach a 50 percent diversion rate from landfill, and it did so without incineration. Thus, we will begin our discussion of Canada with a more detailed look at the Nova Scotia program.

Nova Scotia

I first heard about the Nova Scotia program in 1999, when I was engaged in making the videotape *Target Zero Canada*.[1] This video was shot at a ceremony in Toronto where the organization Earth Day Canada was giving awards to a number of people in government and industry for their innovative leadership in reducing waste; they called these awardees "zero heroes." One of these was Barry Friesen—the waste-resource director for the Nova Scotia government.

At the ceremony Friesen spoke about the province's innovations in sustainable waste management. For me the most exciting thing was hearing that Nova Scotia had built a residual screening facility in front of the Halifax landfill. I had been advocating for such a facility for many years. The argument is fairly basic and obvious. Everyone, including the EPA,[2] admits that all landfills eventually leak, so instead of vainly attempting to control what comes *out* of the landfill, the only sensible way forward is to control what goes *in*. A screening facility built in front of the landfill offers us the best way to do this. This is what Halifax had done, and I couldn't wait to see it with my own eyes.[3] For this beautiful maritime province with its rugged Atlantic coastline the slogan for the program is most appropriate: "Nova Scotia is too good to waste." Maybe we should generalize this slogan to "Our planet is too good to waste."

The History of the Nova Scotia Program

In the early 1990s Halifax tried to expand its landfill. This created a huge uproar from local citizens, who complained bitterly about the terrible odors emanating from the site. Then the municipality proposed a large trash incinerator (750 tons a day) to be built by American company Ogden Martin. Again this produced a huge organized opposition by citizens, and eventually the provincial government rejected the project. At this point the government handed the problem over to the citizens and said, "You don't want landfills and you don't want incinerators. Tell us what you do want. Why don't you design the program?"

The citizens accepted the challenge. The government provided them with all the consultants' reports on which they could base alternative solutions. The government also made no restrictions on their choice except one: The citizens had to reach an agreement by consensus, regardless of how many were involved. The citizens opted for one of the programs listed in the report prepared by Jeffrey Morris of Sound Resource Management Group, a consulting company located in Seattle.[4]

Morris's plan involved source separation and door-to-door collection of recyclables, organics, and residuals; facilities for recycling and composting; and a backup landfill for the residual fraction (Steps 1 through 4 and Step 10 in chapter 2). The citizens made just two changes to the report: 1) Where the report used the word *waste* the citizens changed this to *waste-resources*. 2) Because of their bitter experience with the old landfill the citizens also required that no organic waste go into the landfill without processing.

This second point drove the construction of the residual screening facility in front of the landfill in Halifax. Photographs of the residual screening facility at the Otter Lake Landfill can be seen in figure 2.11 in chapter 2. Figure 9.1 shows more details of this facility.

The residual screening facility is divided into two sections. In the first section (bottom half of figure 9.1) the residual fraction comes into the facility in large plastic bags, and these are dumped onto a tipping floor (see figure 9.2). These bags are then sent to a mechanical cutting device to open them (see figure 9.3) and the contents emptied onto long conveyor belts. A magnet removes iron and steel objects, then the residual fraction is screened by well-protected workers. These workers pull out more recyclables, especially bulky items like cardboard and high-value items like office paper and returnable bottles, which have a deposit on them, as well as toxic items like paint and solvents (figure 9.4).

FIGURE 9.1. An aerial view of the residual screening facility in front of the Otter Lake Landfill near Halifax. *Photograph courtesy of Sun Dancer Ltd.*

FIGURE 9.2. The tipping floor of the residual screening facility. A front-end loader moves the residual fraction onto a conveyor belt for delivery to the bag opener. *Photograph courtesy of David Wimberly*

FIGURE 9.3. This device opens plastic bags containing residual waste with minimum crushing of the contents. *Photograph courtesy of David Wimberly*

FIGURE 9.4. Well-protected workers pull out more recyclables and toxic items from the residuals. *Photograph courtesy of Mirror Nova Scotia*

FIGURE 9.5. A shredder cuts up the dirty organic fraction and other nonrecyclable material (mainly packaging) prior to the biological stabilization operation. *Photograph courtesy of David Wimberly*

FIGURE 9.6. The shredded dirty organic fraction and nonrecyclable packaging materials are moved to long concrete composting troughs. Note the rails on top of the concrete walls. *Photograph courtesy of David Wimberly*

The dirty organic fraction such as kitty litter and diapers is not touched by the workers but continues all the way to the end of the conveyor belts. Here, along with the nonrecyclable items, it goes to the second part of the screening facility. In this section this dirty organic material is shredded (see figure 9.5), and the shredded material is then placed into long troughs with high concrete walls (see figure 9.6). A mechanical turning device—steered by guiding rails on top of the concrete walls—moves through these troughs. This device both aerates the piles, facilitating the composting operation (see figure 9.7), and gradually shifts the material toward the far end of the trough. This aeration and shifting continues for twenty-one days, at which point the stabilized organic fraction exits the trough and is placed in long rows (windrows) in a holding area (see the top of figure 9.1) before delivery to the landfill.

Again it needs to be emphasized that the purpose of this composting operation is not to produce a product for reuse but to stabilize the dirty organic fraction above ground to ensure that when it enters the Otter Lake Landfill the key problems of leachate generation, odors, and methane production are minimized. When I visited the landfill, I could detect no odors and was

FIGURE 9.7. The turning machine moves through the concrete troughs, aerating the compost and shifting it from one end of the trough to the other. The machine is steered through the trough on rails on top of the concrete walls. *Photograph courtesy of David Wimberly*

surprised to see that there were very few birds, indicating little raw organic material remained for them to scavenge (see figure 9.8).

Other features of the Nova Scotia program include:

- Composting facilities for the clean organics.
- Materials recovery facilities for the mixed recyclables.
- Enviro-Depots for the collection and redemption of beverage containers, which have a return/deposit on them (all water, soft drinks, juices, beers, and liquors; only dairy products are exempt from the deposit). Approximately 100 small businesses operate these depots, which also handle a number of household toxics, such as car batteries.
- Drop-off sites for household toxics. The site operating in Halifax is especially convenient. It is open nearly every Saturday, and citizens don't even have to get out of their cars. City employees remove the toxics from the trunk (see figure 9.9) and carefully separate them into different categories, which are then placed in drums (see figure 9.10). These drums are picked up by a hazardous waste company. Nova Scotia also offers a mobile collection system for rural areas.

By operating these household toxic drop-off facilities regularly, with maximum convenience offered to the citizen, the city avoids the long frustrating

FIGURE 9.8. The Otter Lake Landfill receives the biologically stabilized dirty organic fraction from the residual screening facility. *Photograph courtesy of Mirror Nova Scotia*

FIGURE 9.9. City employees unload household toxics from the trunk of a car at a drop-off site in Halifax. Citizens do not even have to get out of their cars. This convenient center is open nearly every Saturday. *Photograph courtesy of David Wimberly*

FIGURE 9.10. City employees sort out household toxics into different categories prior to shipping to a hazardous waste facility. The Halifax center is open nearly every Saturday. *Photograph courtesy of David Wimberly*

FIGURE 9.11. At this site in Halifax vehicles draw up under the canopy of the shelter where the toxics are separated and stored prior to collection. *Photograph courtesy of David Wimberly*

waits often experienced when such collections are offered once or twice a year. Note the relatively short line of cars at this site in Halifax (see figure 9.11).

Renovators Resource

Halifax also sports a very attractive commercial operation that sells reused fixtures from old houses, including large items, such as doors and windows, and smaller items, such as doorknobs and ornamental light fixtures (see figure 9.12). A very creative part of this operation is the furniture that they build from old window frames and other wooden items, including exotic objects like old church pews. One of the owners of this store, Jennifer Corson, had her own TV program in which she showed how old objects could be used in a variety of creative ways in both the garden and in the home. She has published a book on the same subject.[5]

Thousands of Jobs

One thousand jobs were created collecting and treating the discarded materials. These jobs were generated in materials recovery facilities, composting facilities, administration and research, the tire-recycling plant, the waste paint

FIGURE 9.12. Some of the secondhand furniture and bric-a-brack on sale at Renovators Resource in Halifax, Nova Scotia. *Photograph courtesy of David Wimberly*

recovery plant, deconstruction and reuse operations, toxic collection sites, and the environmental depots. In addition another two thousand jobs have been created in the industries reusing the collected material. Nearly all the separated materials are reused in Nova Scotia's own industries.

The program has been the subject of a GPI (genuine progress index) analysis, which unlike other economic indices like the GNP (gross national product) includes estimated social benefits. The GPI analysis was highly positive largely because of the social consequences of generating so many new jobs.[6]

The Nova Scotia program has been an excellent example of cooperation between government officials and local activists.

Nova Scotia's current goal is to bring down the per capita waste generated per person every year from 451 to 300 kilograms by the year 2015.[7]

Ontario

Ontario has had a checkered history in its efforts to solve its waste problems in a sustainable way. Since the mid-1980s the province has provided some good examples, but more recently communities in the Toronto area have shot themselves in the foot by opting for incineration, which from the point of view of sustainability should have been completely eliminated from local decision making. Before I get to the details of this depressing news, let us look at some of the better initiatives at the community level.

Guelph

Located in Western Ontario, this was one of the first communities in Canada to pioneer the wet-dry system. In this two-container system, one container received all the material that could be composted (the wet container or green bag) and the other all the rest (the dry waste or blue bag). The organics went to a composting facility, and a second plant separated out the marketable recyclables from the dry waste. This facility also served the important function of screening all the materials before they were sent to landfill. The program is still going strong, but the city has now added a clear bag for the residual fraction and has also developed a new composting facility.[8, 9]

Georgetown

Georgetown (located about 50 kilometers west of Toronto) has played host to one of Canada's earliest and most successful reuse and repair centers, Wastewise.[10] Wastewise was the brainchild of Rita Landry and her husband, Len, who have both sadly passed away. Rita Landry was one of the most inventive and exciting activists with whom I have had the pleasure to work. She believed in having fun, no matter how serious the struggles became. At the height of a battle to keep a local quarry from becoming a massive landfill for Toronto's trash, she welcomed the premier of Ontario to Halton Hills with a toilet seat around her neck, with a placard that read WELCOME TO HALTON HOLES! Rita related how proud she was when she visited a major newspaper and found the photo pinned up on the notice board in the news room.

Interface Carpet in Belleville

Belleville hosts the multinational carpet manufacturer Interface. When the CEO Ray Anderson read *The Ecology of Commerce* by Paul Hawken,[11] he said, "It was like a spear being pushed into my chest."[12] He vowed there and then to make Interface into the first fully sustainable multinational corporation in the world. He divided the environmental impacts of his company into seven categories: energy use, water use, material use, toxic use, waste production, air pollution, and water pollution. He then recruited all his staff to put their heads together and think of ways to reduce these environmental impacts. He further grabbed their attention by offering to pay them a share in any financial savings the company was able to achieve. Not only were they able to do this, but to everyone's surprise and delight they saved money in all seven categories. This is a great message for other corporations: You can do the right thing and save money doing it. This truly was a win-win situation. By the time Anderson passed away in 2011, his corporate sensibility and environmental ethic was admired the world over.

When touring Interface's Belleville facility, I was truly inspired by the level of industrial responsibility that the workers and managers had been able to achieve there. They had replaced all the toxic inks with a computerized pattern-making operation, and they showed me the used carpet that they were receiving back from customers, which was destined for recycling into new carpet.

The New Democratic Party's Incineration Ban

When the New Democratic Party of Ontario (NDP) was elected into office in 1991, it instigated a ban on new waste incinerators and any expansion of old ones.[13] Many of us were very excited about this—as well as the appointment of many young and enthusiastic recycling coordinators in towns throughout the province. At that time it looked as if Ontario was set to become a model for sustainable waste management, which the rest of North America could copy.

I remember how excited I was when I first walked down a street in Ottawa and saw all those blue boxes neatly lined up in front of every house. Since those giddy days, however, we have learned of the machinations behind this blue box program. It was financed by the beverage industry in an effort to thwart the introduction of bottle deposit legislation in Ontario (Helen Spiegelman provides chapter and verse on this in chapter 18). However, on the plus side Ontario has the most successful and efficient deposit system for the collection of reusable beer bottles I have seen in any country. This is operated by the industry itself (The Beer Store) and imposes no burden on the community rate-payers at all.[14] No wonder the soft drinks industry was worried!

It's also no wonder that when the NDP was voted out of office one of the first things that the incoming conservative party did was to lift the ban on incineration.[15] The incineration industry (particularly Covanta Energy) had kept front men in Ontario for such an eventuality, and at every turn they have done their level best to persuade politicians to go for incineration instead of more intensive recycling and composting efforts.[16] Of course, they never couch their arguments in these terms. They always argue on the basis of there only being two choices: landfills or incinerators. They agree that recycling and composting are important but give little credence to the notion that they can compete with either landfilling or incineration for a major portion of the waste stream. Despite the lifting of this ban, it took sixteen years before an incinerator was approved.

On the bright side, the influence of the recycling coordinators is still being felt throughout Ontario today. The blue box curbside recycling program carries on, emphatically demonstrating that citizens were not the problem: They took this source separation in stride.

Trenton

One of the most successful towns in the curbside collection of discards was Trenton (part of the Quinte region), which reached a diversion rate of 70 percent by the early 1990s. In addition to the blue box program and the building of an MRF, the town provides householders free containers for backyard composting of kitchen waste, drop-off sites for household hazardous waste, and a pay-as-you-throw (PAYT) system for the residual fraction. Many people were pessimistic about such a financial penalty and thought that some citizens would dump their waste over the nearest fence (fly-tipping). The organizers studied this matter carefully and found that the PAYT system did not increase fly-tipping. Their explanation for this unexpected result was that citizens were presented with so many options in town for getting rid of their discards legitimately that there really was no need for people to go through the extra effort of finding somewhere else to get rid of it illegally.[17]

When I visited the Trenton operation in the late 1990s and asked the organizers how they got such great compliance, they informed me that they told citizens that to landfill a ton of waste cost the city $150, but to recycle it cost only $100 a ton, so that each ton of waste that citizens helped to recycle saved the town's taxpayers money.

Toronto

The city of Toronto raised the expectations of the zero waste movement worldwide when, following San Francisco's lead, they declared a zero waste strategy in January 2008.[18] For a time the city appeared to be quite serious about this goal. It already had a blue box program, and in addition it encouraged backyard composting and then added a green bin for the separate collection of clean organics. For these it built anaerobic digestion facilities. Things looked very promising until Covanta Energy was able to persuade the York-Durham region to host a massive incinerator.[19]

Ottawa

For thirty years, my wife and I lived approximately eighty minutes by car from Ottawa, which is one of our favorite cities. We frequently went there for shopping, movies, restaurants, and the wonderful National Gallery. Each time we visit we are impressed anew with the livability of this city. In winter some people skate to work on the extensive canal system, and in summer the same system sports canoes, sightseeing boats, and joggers along the paths that border the canals. To this we can add the unusual black squirrels and the wonderful displays of tulips donated by the queen of the Netherlands in gratitude for her stay in this city during World War II.

As far as waste management is concerned Ottawa was one of the first of Ontario's cities to embrace the blue box system for the curbside collection of recyclables and also instigated an even more revolutionary "Take It Back!" program in which retail stores take back used paints, solvents, batteries, electronic goods, ink cartridges, and pharmaceuticals.[20] Indeed, Ottawa looked set to move in a serious zero waste direction until the mayor and council got involved in a proposal to build an experimental plasma arc incinerator in the city (see chapter 3 for more discussion on this project).

Edmonton, Alberta

Supporters of the Edmonton waste program applaud the diversion from landfill (over 60 percent) that it has been able to achieve. However, I had mixed feelings about what I saw when I had a chance to visit their gigantic site for recycling and composting in the early 2000s.

At Edmonton's waste management center, Waste Management of Canada operates an impressive MRF that handles two different streams of recyclables collected curbside—one stream consists of all the paper products and the other all the bottles and cans. However, the city does not have curbside collection of clean organics. Instead, these go into the residual container. These residuals end up going to a huge mixed waste composting facility (the largest composting operation under a single roof in North America) operated at this site. I dread to think where the 80,000 tons per year of dirty "compost" ends up.[21]

Another disappointing aspect of this Edmonton operation is that despite the education and research buildings available at their waste management center, little effort has been made to develop a zero waste research center to study the residual fraction. Instead, the Alberta government has provided a subsidy to help Edmonton build a gasification plant on the site to convert the bio-organic fraction in the residual fraction into ethanol.[22]

Vancouver, British Columbia

Vancouver also seems determined to shoot itself in the foot. Despite declaring a zero waste policy, it too is proposing more incineration in the Vancouver area (it already has a Covanta Energy–operated incinerator in the suburb of Burnaby). In fact, the metro government has discussed building as many as four more incinerators—much to the consternation of many environmental activists in the area.[23] This is particularly upsetting because Vancouver has taken the lead in terms of local industries developing ambitious programs for extended producer responsibility (EPR) for their packaging and some other products.

Vancouver is also home to Helen Spiegelman, who has been one of the most articulate proponents of EPR and is now horrified at her city's push to build incinerators.[24] After all, the notion behind EPR is to transfer responsibility to the manufacturer at the front-end, where packages and products are made. Incineration returns the issue again to the back-end of the problem, after these materials have been thrown away. By building incinerators communities will pay a fortune to allow manufacturers to escape responsibility for badly designed packages and products. These resources will simply go up in smoke.

The situation in Vancouver prompts the question of how serious people can be about zero waste when on the one hand they say they support zero waste and on the other they plan to build massive incinerators. The answer (or rather deception), as we've learned, comes in the description "zero waste to landfill." For those who have worked so hard promoting genuine zero waste it is disheartening to see the concept corrupted in this way, especially by people who pride themselves on their progressive attitudes. Genuine zero waste means "zero waste to landfills *and* incinerators" or "zero waste disposal." For those genuinely interested in moving toward sustainability these latter goals are really worth struggling for. "Zero waste to landfill" is "business as usual" for those who manufacture throwaway materials that are hard to recycle, as well as the incinerator industry that makes them "disappear" in smoke.

What we have here and elsewhere in Canada is a war between waste management in the community interest (zero waste) and waste management in the corporate interest (megalandfills and incinerators). This struggle is illustrated in several essays in Part 3 of this book.

Even the much-lauded EPR program in British Columbia has not lived up to expectations, partly because of the subversion of allowing incineration in their mix of options. In 2011 Nadine Souto reviewed the British Columbia EPR program for the Institute for Local Self-Reliance (ILSR). Based on her review and input from Daniel Knapp, zero waste advocate Neil Seldman had this to say about the program:

> British Columbia's EPR enabling law puts resource destruction on an equal footing with resource conservation. It explicitly endorses "waste" incineration, placing it number three on a list of four alternatives for discard management. In fact, burning resources shares 3rd place with "recover material . . . from the product," just above the 4th and last resort, landfilling. Sensing a big opening, incinerator vendors have flocked to the province and are now forcing existing recyclers, EPR advocates, and downwind communities to mobilize against what they see as subsidized competition for the resource flows.[25]

There is also concern that single-stream collection of recyclables could replace multistream source separation. According to Clarissa Morawski, a waste expert at the Container Recycling Institute, single-stream collection systems

- blend materials that should be kept separate;
- downgrade commodities like paper, plastics, glass, and even aluminum;
- disrupt markets by driving up costs for remanufacturers who seek quality feedstocks;
- depress prices paid by resource brokers and traders for "upgraded" recyclables coming out of single-stream MRFs because of much higher unrecyclable "residuals"; and
- lead either to the incineration of high tonnages of EPR materials or to landfilling them.[26]

In this ILSR report, Seldman et al. also cite evidence that some packaging companies are using the current situation to get rid of current deposit systems, which according to Morawski are far superior to other methods of recovering beverage containers because container deposit-return (CDR) systems

- create 11 to 38 times more jobs than curbside recycling systems relative to beverage containers;
- generate dramatically higher volumes of beverage containers than curbside systems, an average of 76 percent recovery in CDR states compared to just 24 percent recovery in non-CDR states;

• mean that glass bottles have six times more recycled content than bottles made in a state without a container deposit (72 percent versus 12 percent).[27]

Say Seldman and his report coauthors, "If you want more jobs, choose deposits; if you want more resources, choose deposits; if you want to close the loop to prevent mining, choose deposits."[28]

The ILSR's recommendations on how British Columbia could improve its EPR program included

• strengthening the existing recycling networks and nodes by arranging to pay them fair market prices for their disposal services;
• removing incineration from the hierarchy;
• banning the composting of mixed municipal solid waste;
• restoring and enhancing source separation by allowing all sorts of fees-for-service to pay operating costs plus profit for niche recovery enterprises that find higher and better uses for all discard categories;
• building new centralized zero waste transfer depots laid out like airports and making them into places where responsibility and ownership of all discarded materials can change hands legally, pleasantly, and profitably;
• adopting policies favoring specialist enterprises within these structures and allowing for growth and differentiation of the industry, including remanufacturing.[29]

Nanaimo, Vancouver Island

Next door to Vancouver on Vancouver Island, the Regional District of Nanaimo (RDN) provides positive news. In 2002 the district committed to zero waste as its long-term waste reduction and diversion target, and unlike Vancouver seems to mean what it says.

According to the district's website, "zero waste focuses on reducing the region's environmental footprint by minimizing the amount of waste that must be landfilled through reduction, reuse, recycling, redesign, composting, and other actions. The RDN was the first jurisdiction on Vancouver Island and one of several forward looking local governments in Canada and around the world to move beyond recycling and adopt a zero waste approach to eliminating waste."[30]

In an effort to curb the amount of waste headed for landfills, Nanaimo introduced Canada's first PAYT residential garbage collection system in 1991. After that, it expanded curbside recycling programs; banned paper, metal, commercial food waste, clean wood waste, and other recyclable materials from the landfill; and promoted composting.[31] Nanaimo has a number of notable milestones in its movement toward zero waste.

1989: Residents and businesses divert 10 percent of solid waste from the landfill.

1991: The RDN introduces Canada's first PAYT residential garbage collection system.

1995: Reduction, reuse, and recycling initiatives divert 26 percent of solid waste from the landfill.

2000: The RDN and its municipal partners divert 54 percent of the total waste generated in the region—exceeding the 50 percent target set by the provincial government.

2002: The RDN adopts zero waste as its long-term waste diversion target.

2004: The RDN prepares an updated solid waste management plan, which sets an interim goal of diverting 75 percent of the region's waste from the landfill by 2010.

2005: The RDN bans commercial food waste from the landfill. A commercial food waste diversion program involving businesses and organizations diverts more than six thousand tons of food waste and organic compostables annually from the landfill.

2007: The RDN and its municipal partners launch a residential food waste collection pilot project that will provide the information needed to develop a regionwide program.[32]

Before we get too excited about this program it should be noted that every time the authors mention zero waste it is coupled with "to landfills," so one wonders whether they are including material sent to an incinerator (Burnaby, perhaps) as part of their diversion rate.

The Gibsons Resource Recovery Center

A ray of hope for genuine zero waste efforts in British Columbia is coming from the small town of Gibsons, located on the coast. There you will find an operation called the Gibsons Resource Recovery Center, run by a fascinating entrepreneur called Buddy Boyd and his partner, Barb Hetherington.

Their facility has never received any funding from taxpayers. Instead, they generate their own revenue from the fees they charge for accepting certain materials, as well as from the sale of clean recyclables and the sale of other objects they mine from the discard stream that comes through their gates. The facility currently employs thirteen people.[33]

The center accepts all kinds of household discards and used items to be resold or remade into new products. In all they accept twenty-one different items ranging from textiles to electronic equipment.[34, 35] Some items are received without a fee; others have a fee charged. Boyd advocates a PAYT system, especially for things that are hard to recycle. He says, "It should not be a free service, because it detaches people from responsibility for the things that they buy."[36] According to a newspaper article, "Boyd sees garbage differently than most. He describes what he does as 'reclaiming garbage and putting it back into the community, not the ground.'"[37] It sickens Buddy that while he struggles daily to conserve all the resources he can out of the local discard stream with no help from the British Columbia government, this same entity is planning to fritter away millions (billions?) of dollars on new incinerator projects for the Vancouver area while claiming that it is supporting zero waste (see chapter 17)!

Beyond Italy:
Other Initiatives in Europe

Italy, as we saw in chapter 7, is among the world's zero waste leaders. But not all of Europe is so forward-thinking when it comes to handling its waste in a sustainable fashion. Countries in northern Europe have a huge problem moving toward zero waste because most have built their waste management systems around incineration. Denmark and Sweden have compounded their inflexibility by building their district water systems around the waste heat generated by these incinerators. So it would be very difficult for them to shut down these facilities—and impossible for them to get to zero waste, since they have to feed so much waste into these burners to keep them operating.

In some countries—like the Netherlands, Germany, and Sweden—as citizens have gotten better and better at recycling and composting, incinerator operators have been forced to import waste from other countries to keep their facilities going. Currently, Germany recycles 62 percent of its waste. One wonders what the rate would have been if Germany hadn't built so many incinerators.

As long ago as 1986, Heidelberg, Germany, had an advanced recycling system. And in 1991 German citizens opposed to incineration in Bavaria narrowly missed passing a law, Das Bessere Müllkonzept (the Better Waste Concept), that embodied many of the ten steps to zero waste outlined in chapter 2. So if the mainland of Europe were to ever throw off the shackles of incineration, I suspect that many communities would be able to move quickly toward zero waste, a key stepping-stone to sustainability.

Still, we can find many good examples of European communities pushing for key components of a zero waste strategy, such as source separation, reuse, recycling, repair, composting, waste reduction initiatives, PAYT systems, and EPR regulations. Even in countries with a heavy commitment to incineration, there are communities taking positive steps.

The goal now is to urge more communities in more European countries, even those relying on incineration, to get creative and adopt zero waste strategies. It is tantalizing to imagine the kind of residual screening and research facility German or Scandinavian engineers and designers might build in front of a landfill to handle the residual fraction if they could no longer burn it.

Flanders, Belgium

Flanders is the Flemish-speaking region of Belgium's three regions. It is made up of five provinces that have a total of 308 municipalities and 6.2 million inhabitants. It has developed a remarkable program using every device and creative incentive anyone has ever thought up to minimize the amount of waste going to landfills and incinerators and to maximize the amount of materials recovered. Moreover, Flanders has managed to stabilize residential waste *production* since 2000 through waste prevention programs.

The results are impressive. While Flanders still burns 24 percent of its waste, it now has the highest regional waste diversion rate in Europe, with 73 percent of the residential waste produced in the region getting reused, recycled, composted, or otherwise diverted from incineration. Had decision makers in Flanders known in the 1980s and 1990s that it could achieve such results it is questionable whether they would have invested as much money to build incinerators as they did. Instead, like San Francisco, they might have pushed for a zero waste strategy without using any incineration at all. Nevertheless, what Flanders has achieved is remarkable for the size of population involved and provides a model for the rest of the world. My major caveat for decision makers elsewhere: Copy Flanders on their innovative programs to reduce, reuse, recycle, and compost their waste but do so without building incinerators.

The most progressive thrust of the program of the Flanders regional waste authority, called OVAM, is aimed at industry and encourages less wasteful product design. In a GAIA report, Cecilia Allen outlines its major features:

The "Ecolizer": a tool for designers to estimate the environmental impact of products. It includes a set of environmental impact indicators relating to materials, processing, transport, energy, and waste treatment, allowing designers to identify opportunities to reduce those impacts by changing the design. For instance, one can calculate the environmental burden of a coffee machine by finding scores for different indicators—the materials, the manufacturing process, the related transport, and the treatment after the product is discarded—and then evaluate options to reduce the environmental burden score by making changes in the design of the coffee machine.

Ecoefficiency Assessment: a program to evaluate the ecoefficiency of small- and medium-sized companies. It identifies points of intervention to reduce waste, improve energy and water efficiency, increase recycling, and so on. The test is free of charge. OVAM consultants follow up to implement the changes. As of 2009, one thousand companies had been assessed.

MAMBO: a software program that allows companies to calculate the direct and indirect costs associated with waste, including those resulting from waste treatment and inefficiency.

Inspirational Online Database: a collection of case studies of businesses that have implemented clean production and ecodesign methods.

Ecodesign Awards: designed for students and professionals as a way to encourage innovations in waste prevention. The prizes range from $508 to $5,080.[1]

OVAM also has strong programs, including substantial subsidies, to encourage reuse and recycling. Consequently, Flanders has more than one hundred secondhand shops that resell items that would have otherwise hit the waste stream. According to Allen's report, these shops employed 3,861 people and had more than 3.6 million paying customers in 2009.

Equally important, notes Allen, is the Flanders focus on reducing household waste by incentivizing composting:

> In Flanders, successful approaches have included annual charges for the collection of organic materials ($51 for a 120 liter bin), educating citizens about home composting through communication campaigns, promoting "cycle gardening" to reuse yard waste, encouraging composting at schools, and composting demonstrations at community compost plants. A "compost masters" program has also been established, through which citizens are trained in composting and then encouraged to work as volunteers training other citizens and assisting them to compost properly. By 2008, 4,000 citizens had been trained, and there were 2,500 active master composters. These efforts have yielded significant results: it is estimated that about 100,000 tons of organic materials were kept out of the collection and management system in 2008, thanks to home composting. In densely populated areas, the government encourages community compost plants, where citizens can take their organic materials. These facilities usually use compost bins, and so do not take up much space. The success of this program continues to grow. By 2010, approximately 34 percent of the Flemish population—almost two million people—was composting at home.[2]

All in all the Flemish program has been both creative and effective. One original touch, which I found both creative and amusing, is the fact that the

Flemish authorities provide free chickens to consume some of the kitchen scraps not dealt with by composting.

Basque Country, Spain

At one of the many zero waste conferences that have taken place in Capannori in Tuscany, Italy, I was surprised and delighted to hear of a novel way of collecting source-separated materials. The mayor of Ursibil in Basque country in Spain modestly calls this system the "Italian method" because it was modeled after the Capannori system of using the same trucks to collect different source-separated materials on different days of the week. However, Ursibil had added a novel twist. Instead of householders putting the containers on the sidewalk, they put them on special numbered hooks either on walls or posts.

The same system has been instigated in the town of Hernani (population 19,000, located nine kilometers from San Sebastián) and is now spreading throughout the Basque country (see figure 10.1). Each household has their own numbered hook and four correspondingly numbered and color-coded containers, each for a certain type of waste. Pickups occur six days a week. The organic fraction (in a brown container) is picked up three times a week. This not only allows citizens to separate their waste more thoroughly, it accommodates frequent pickups without having a constant stream of curbside bins. It also offers residents both transparency (in relation to the municipality) and privacy (in relation to neighbors). Only the municipality has a list of which number belongs to which household.

The communities using this system are getting very high diversion rates (over 80 percent) very quickly. Ursibil itself reached a diversion rate of 82 percent in a few months after starting the program. According to GAIA, in 2008, before door-to-door collection started, Ursibil was taking 175 tons per month to the landfill. One year later, the amount had dropped to 25 tons.[3]

Once again the enthusiasm of both citizens and politicians for this system has come from a strong desire to keep incinerators out of the region. In 2002, when the San Marko Landfill—used by Ursibil, Hernani, and other local municipalities for most of their waste—was nearly full, the provincial government presented an emergency plan with two components: one bad (a proposal to build two incinerators) and one good (the introduction of door-to-door collection of compostable waste).

There have been huge demonstrations held in opposition to the two incinerators (see figure 10.2), but the source separation of the compostable fraction has been accepted as a rational response to the problem of reducing the flow of materials to the landfill.

FIGURE 10.1. A wall in Hernani holds the numbered hooks on which neighbors place their containers for separated discarded materials. If walls are not readily available, specially constructed posts are used instead. *Photograph by Gipuzkoa Zero Zabor*

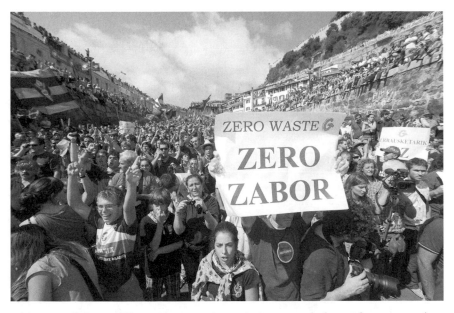

FIGURE 10.2. Citizens in Hernani protest against an incinerator and advocate for zero waste (*zero zabor*). *Photograph by Gipuzkoa Zero Zabor*

In reality, the fact that both the citizens and political leaders have worked together and embraced the positive and concrete alternative of a comprehensive source-separation system has been their most effective weapon in combating incineration. Already one incinerator has been defeated. Construction of the other has started but is currently on hold. Meanwhile, more and more politicians opposed to incineration have been elected, so it is highly unlikely that this second incinerator will ever operate. Said the mayor of Hernani: "Our state-of-the-art technology is the neighbors. If the neighbors separate well, there is no need to build an incinerator."[4] See the sidebar on Hernani for the practical details of the program.

Meanwhile, there are efforts to get the whole region to adopt zero waste, and the enthusiasm runs strong. When I attended a conference on zero waste in nearby San Sebastián in 2012, posters advertising the conference adorned nearly every lamppost in the city. I was also impressed to see several community composting projects in Ursibil. Each served between eight and sixteen householders and was located on a site in parks adjacent to users' apartment buildings—in some ways this is similar to the community composting efforts underway in Zurich, Switzerland.[5]

But the most impressive aspect of the whole history and the operations in both Hernani and Ursibil is the enthusiasm of the local officials for the program and their willingness to work closely with activists and citizens in general. In a presentation at our zero waste conference in San Sebastián[6] the former mayor of Ursibil, a nationally famous sports star now in the national parliament who first introduced the program there, said that in a questionnaire citizens were asked if they were *proud* of their program. He said he was thrilled and moved when over 80 percent replied yes.

Catalonia, Spain

GAIA's Joan Marc Simon, who heads up Zero Waste Europe, shares his time between his native Barcelona, in Catalonia, and Brussels. In Catalonia, he says, "More than one hundred municipalities have implemented door-to-door separate collection, and the law requires separate collection of organics in all municipalities." He also notes that the region created a forum for the Catalan Zero Waste Strategy in 2011 to change its paradigm from one based on waste disposal to one based on zero waste, and to build upon four zero waste pillars: a network of zero waste municipalities, zero waste organizations, zero waste companies, and zero waste universities.

The support for this strategy has come rapidly, says Simon. "In only one year of existence six municipalities have adopted zero waste goals; nine

Key Features of the Hernani Program

BY CECILIA ALLEN

The Collection System

Each stream has a designated pickup day: organics on Wednesdays, Fridays, and Sundays; light packaging on Mondays and Thursdays; paper and cardboard on Tuesdays; and residuals on Saturdays. Light packaging is placed in bags, and the government sells reusable bags for this purpose. Paper and cardboard are tied in bundles or placed in boxes or bags. Organics are placed in the bins provided by the government, and the residuals are disposed of in bags. Collection is done at night, with a complementary shift during the morning.

A Separate System for Glass

For glass, the system of large containers on the streets was maintained, and door-to-door collection is done only in the old part of the city. A nonprofit association created by producers, packers, bottlers, and recyclers handles this stream. The association is funded by contributions the packaging companies pay for each product they put on the market.

Emergency Backup Centers

If someone misses the door-to-door collection, there are four emergency centers to drop off waste. There is also a drop-off site that takes bulky waste, electric and electronic devices, and other waste not covered by the door-to-door collection—free of charge. For businesses, the collection schedule is the same as for households, with an extra day of collection for residuals.

Monitoring the System

If the collector identifies a stream that does not correspond to that collection day, she or he puts a sticker with a red cross on the bin and does not collect that waste. The information is given to the administration office, and the household receives a notice explaining why the waste was not collected.

Home Composting

Hernani promotes home composting throughout the municipality. People can sign up for a composting class, request a home composting manual, and receive a compost bin for free. There is a phone line to get composting advice, and there are compost specialists who can visit households in need of assistance. People who sign up to compost at home receive a 40 percent discount on the municipal waste management fee.

The Fee for Businesses

The fee for businesses varies according to the collection frequency and the amount of waste produced, using PAYT criteria.[7]

universities have embraced zero waste; a growing group of organizations are gathering around the strategy; and a working group of zero waste companies from different sectors—from the printing industry to copper recycling—meet periodically to exchange best practices."

The Catalan Zero Waste Strategy calls for the region to reduce waste generation by 20 percent and reverse the current situation of 30 percent recycling and 70 percent disposal to 70 percent recycling and only 30 percent disposal by 2020.[8]

Gothenburg, Sweden

As far as the environment is concerned Sweden is one of the most progressive countries in the world: It was the first country to ban mercury amalgams, and it rejected water fluoridation in the 1970s. You will recall that Sweden embraced incinerators early on and even tied its district water-heating systems to incinerator-generated energy. Swedish companies have also marketed their incinerator technology to other countries. But to its credit, Sweden was also the first country to put a moratorium on building new incinerators in 1985 when it became clear that they produced dioxins, which were accumulating in cows' milk near the facilities and in human breast milk. When they lifted the moratorium in 1986, it was with the understanding that new incinerators would have to meet a tough new dioxin emission guideline. This Swedish guideline was adopted by Germany as its standard and eventually by the entire European Union. Whether or not all the European incinerators meet these standards on a routine basis is another matter.

With so many incinerators to feed it is of little surprise that Sweden has not blazed a trail to zero waste. However, there are some things that it has done, which are very positive and could be copied and incorporated into zero waste programs elsewhere. In fact, one of the most creative antidotes to waste management's more boring solutions—landfills and incineration—can be found in Sweden's Kretsloppsparken, a reuse and recycling park in Gothenburg.

Here are just some of the creative ways that Paul Martensson, of the city's Department of Sustainable Water and Waste Management, attracts the public and especially children to use this reuse and recycling facility and have fun doing so (see figures 10.3 and 10.4).

The 330,000-square-foot facility was built largely with reused building materials. All visitors are met in a large hall, which Martensson calls the "marriage point" between the park operators and the public. Here visitors are greeted with a live band playing music and are persuaded to donate all kinds of secondary materials to the park.

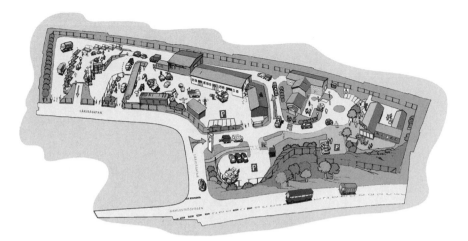

FIGURE 10.3. An artist's schematic representation of Kretsloppsparken, the reuse and recycling park in Gothenburg, Sweden. *Illustration courtesy of Kretsloppsparken*

FIGURE 10.4. The reuse and recycling park in Gothenburg occupies 330,000 square feet. In the foreground are drop-off containers that receive many different categories of recyclables. Immediately behind these containers are a series of buildings used for visitor reception, flea markets for reusable items, repair shops, art galleries, a restaurant, and entertainment areas for children. *Photograph courtesy of Paul Martensson*

Once inside the park, children are greeted by clowns (see figure 10.5) and entertained by jumping castles. Visitors can view many paintings, photos, and sculptures that have been donated to the park. Some of the sculptures have been made out of discarded materials. The walls inside one of the bathrooms at the facility make an amusing venue for some of the artwork; the novel viewing area has actually increased sales.

For teenagers the park organizes rock concerts (see figure 10.6). A twist here is that the entrance fee is paid with collected discards—perhaps 20 plastic bottles. Again, the message is all about helping to save the planet but having fun doing it.

Famous people are persuaded to donate an item that they have owned. Even a sock or shirt from a celebrity could bring in a nice resale fee. In addition, local TV personalities who specialize in antiques come to the park to give advice to visitors on the value—or otherwise—of their collectibles (see figure 10.7).

But perhaps the most entertaining part of the operation is the famous sorting dog (see figure 10.8), which has been trained to separate a mixture of discarded materials into six different fractions. I joke about this in my zero waste presentations by suggesting that if a dog in Sweden can separate waste into six different categories surely citizens elsewhere can separate into three or four!

FIGURE 10.5. One of the clowns that entertain children at the reuse and recycling park. *Photograph courtesy of Paul Martensson*

FIGURE 10.6. The price of admission at a rock concert at the reuse and recycling park—waste currency, as in twenty plastic bottles. *Photograph courtesy of Paul Martensson*

FIGURE 10.7. A famous auctioneer from a televised antique show appraises the public's wares. *Photograph courtesy of Paul Martensson*

The park also holds a fashion week during which young models sport outfits made from secondhand apparel, making these clothes fashionable rather than inferior in young people's minds. Used clothes are arranged carefully.

Throughout the park—in what amounts to a giant fleamarket—dozens of "salespeople" sell secondhand items in stalls. Figure 10.9 shows a well-organized layout for various recovered building materials.

Other activities in the park include green farmers who come in to allow people to taste and buy their organically grown food; an exhibition of "environmental cars," where people can view and test drive these ecofriendly vehicles on the ecodriving course; running and bike competitions; a scrap-sculpture exhibition and competition; and a recycling day featuring marching bands. There are even workshops for people who want to learn how to repair their old possessions rather than discard them (see figure 10.10).

Martensson has spoken about this park at zero waste conferences held in Brazil, San Diego, San Francisco, Italy, Spain, and Bahrain in recent years and is a one-man global whirlwind celebrating the "fun message" of both Kretsloppsparken and the potential of zero waste. At the 2012 United Nations Conference on Sustainable Development in Rio de Janeiro, Martensson even presented the king of Sweden with a T-shirt that translates to "I love

FIGURE 10.8. This famous sorting dog has been trained to separate materials into six different categories. *Photograph courtesy of Paul Martensson*

FIGURE 10.9. Many building materials and other reusable items from old houses can be purchased at the reuse and recycling park in Gothenburg, Sweden. *Photographs courtesy of Paul Martensson*

FIGURE 10.10. The bicycle repair section in the reuse and recycling park in Gothenburg, Sweden. *Photograph courtesy of Paul Martensson*

Kretsloppsparken" and urged him to wear it when he was mowing the lawn around his castle (see figure 10.11).[9]

United Kingdom

The government in the United Kingdom has long been trapped between the false dichotomy of landfilling and incineration. Only relatively recently have communities in the United Kingdom started to achieve over 50 percent diversion rates with source separation of recyclables and compostables. Even so, many counties are still opting to build hugely expensive incinerators to "solve" their waste problem. A particularly depressing example has occurred in Gloucestershire.

The city of Gloucester had designated Javelin Park (the site of an old airplane manufacturing facility) on the outskirts of the city for industries offering to take care of the residual fraction. While citizens were enthusiastic about this being used for a zero waste resource recovery park the local council has opted to build a trash incinerator there instead.[10]

Because I grew up in England and still have many friends and relatives living there, throughout my twenty-eight-year involvement with waste issues I have frequently returned to my home country to help communities whenever I have had the opportunity to do so. The experience has been mixed. On the

FIGURE 10.11. Paul Martensson (left) presenting the king of Sweden at the recent Rio de Janeiro conference with a T-shirt advertising his reuse and recycling park, bearing the legend I LOVE KRETSLOPPSPARKEN. *Photograph courtesy of Paul Martensson*

one hand I have been very well received at the grassroots level both in efforts to fight off incinerator proposals and more recently in the promotion of zero waste. However, at the governmental level—especially at the national level—until very recently there has been little interest in what I have had to say. My first exposure to this governmental disdain came very early on.

In 1986 I visited the UK Department of the Environment on my return from a visit to Germany, where I had visited and recorded source-separation programs, including Heidelberg's three-container system for curbside collection—one for compostables, one for recyclables, and one for residuals (the very same system now practiced by many communities around the world).[11] I was received very politely by the UK department official, who proceeded to tell me that I need not worry about Britain building any more incinerators because 1) they couldn't afford them and 2) there were too many holes in the ground they wished to fill up! I responded that, before the country went back to landfilling all of its waste, it might want to consider some of the excellent source separation, recycling, and composting programs in Germany. Before I could give any details I was interrupted by my host, who said, "Quite frankly, Dr. Connett, the Germans have ways of making people do things that we just don't do over here in Great Britain!" This experience was more like a sketch from Monty Python than a serious discussion with a regulatory agency.

It is a huge shame in my view that UK government officials didn't take inspiration from Germany those twenty-seven years ago. In the interim years a huge amount of time and money has been wasted doing the wrong things—including building incinerators that they couldn't afford.

County Councils versus Local Councils

In the United Kingdom municipalities are responsible for the collection of waste and the counties are responsible for the waste handling facilities such as landfills, incinerators, and recycling facilities. Over the years lip service has been paid to recycling while landfills have continued to dominate disposal. Each new incoming government has tried to drum up support for new incinerators (many were closed in the late 1980s because of high dioxin emissions). While some municipalities have offered curbside recycling over the years, not many until recently have provided curbside collection of organics.

The Landfill Tax

The national government was finally forced to act by European laws to dramatically limit landfilling. This precipitated a surcharge of about $12 per ton for each ton of waste delivered to the landfill, rising by the same amount each year. By 2013 this surcharge had risen to a massive £64 ($96) per ton. Currently, the whole of this surcharge goes to the national government and disappears into the general exchequer with very little, if any, of it going back to the counties to support reuse, recycling, or composting measures.

The results of this surcharge have been both positive and negative. For the more progressive counties the surcharge has been used to propel source separation, recycling, and composting, with many achieving diversion rates of around 55 percent by 2011.[12] Clearly, anything diverted from landfill saves money. But other counties have been panicked into building very costly incinerators. For example, Suffolk is considering building an incinerator with lifetime costs of over £1 billion.

Ironically, with the escalating landfill surcharge in place the most expensive option (incineration) has been portrayed as a cost-saving measure. However, these huge capital outlays have been partially hidden by private finance initiative (PFI) schemes. Taxpayers usually don't find out about the huge debt that has been incurred by the county until it's too late—after long-term "put or pay" contracts have been signed by gullible, or possibly corrupt, politicians.

These hugely irresponsible economic investments are propelled by the false notion that the only two things you can do with waste are bury it or burn it. All too frequently only lip service is paid to recycling. Nowhere is this clearer than the manipulation of the waste hierarchy. The waste hierarchy, which is generally accepted in Europe, lists waste management methods in order of priority as far as the environment and society are concerned:

- Reduction
- Reuse
- Recycling (which includes composting)
- Energy recovery (which usually means incineration, although it could include anaerobic digestion)
- Landfilling

Incinerator promoters have used this hierarchy as a respectable rhetorical cover for moving from landfills to incinerators—moving from the least-desirable option to the next-least-desirable option—and calling that progress. I argue that we should use financial incentives to place the emphasis on the top of the hierarchy (reduction and prevention) and then move down through reuse, recycling, and composting.

Incentives for Moving from the Top of the Waste Hierarchy Down

In 2010 I had a meeting with the waste section of the UK Department for Environment, Food, and Rural Affairs (DEFRA) and recommended to them that they change their landfill surcharge system to a surcharge *and rebate* system. In other words, penalize the bad (landfills and incinerators) and reward the good (reduction, reuse, recycling, and composting). So in a system that could be made fiscally neutral, surcharges would flow from the counties to the central government for each ton of waste burned or buried (including incinerator ash), and rebates would flow from the central government to the counties for each ton that was reduced, reused, recycled, or composted. I argued that this would drive the waste hierarchy from the top down rather than from the bottom up.

Sadly, the DEFRA officials showed little interest in this rational approach to driving the waste hierarchy in the right direction. Instead, they seemed far too attached to keeping incineration in the mix.

Despite the gloomy vision from those who are paid large salaries to protect the environment, there are still shining examples of initiatives taking place in the UK at the community level. I heard of one just recently: Refurnish in Doncaster. This reuse and repair operation was inspired by Urban Ore in Berkeley. Refurnish picks up bulky waste, runs a store where scrap is sold, and offers a handyman service that incorporates reused items. Refurnish director Jim McLaughlin says the operation also employs people

> to assemble and repair wooden furniture and soft furnishings, and has just started experimenting with cutting carpeting into tiles for reuse. But most importantly, we focus on making an impact on social issues through affordability of goods and employment/ training programs.[13]

An Exciting Development in Coventry

An exciting development on the zero waste front has occurred in Coventry. This city, while hosting an old incinerator, is offering support for an initiative from Coventry University and the local chapter of Friends of the Earth to set up a zero waste research center.[14] This is part of the crucial Step 8 in the ten steps to zero waste outlined in chapter 2. The city is offering facilities where a team of university professors and students can intercept the residual fraction going to the incinerator and study it. Those involved are reaching out to researchers elsewhere, including Capannori in Italy and the operation Cwm Harry in Wales.

The Cwm Harry Program

BY KATY ANDERSON

The mayor of Presteigne and Norton, James Tennant-Eyles, confesses that at first he thought the idea presented to him by Cwm Harry wouldn't work. "About two years ago," the mayor explains, "Adam Kennerley, the CEO of Cwm Harry, came to ask me whether the community of Presteigne and Norton would be up for a zero waste project. Adam talked about this new way of waste collection involving 'slow recycling'—going round building relationships with the community and sorting waste on the curbside, and I had thought it all seemed quite barmy, but two years down the line it has all come to pass, and very successfully so."

Today, the Zero Waste Service is one of three projects funded by the Welsh government to find ways of meeting its challenging zero waste targets: 1.7 percent reduction per year and 70 percent recycling by 2025.

The Zero Waste Service uses a slow recycling approach, similar to the idea of slow food. Slow recycling works on a human scale, prioritizing people and high-quality materials. Cwm Harry provides a curbside sort into fifteen streams for 250 homes collected in a converted milk float formerly used for milk home delivery (see figure 10.12); composts the collected food waste (see figure 10.13); provides refuse collection in a Ford Transit van with a caged tipper; and runs a farm-based depot to store materials for resale.

FIGURE 10.12. A slow-recycling worker separates fifteen streams of recyclables in a collection vehicle made from an old electric-powered milk float. *Photograph courtesy of Cwm Harry*

Cwm Harry also runs a series of events to get people thinking about waste and provides advice to families to help them recycle more. In this way the whole town looks after all of its discarded materials.

This approach has led to a steep change in attitudes toward waste, breaking through the Welsh targets to achieve a 24 percent reduction in waste and 75 percent recycling. Three factors have led to this change.

1. **Making Waste Visible.** Rather than whisking waste away early in the morning in a big truck as most conventional schemes do, the Zero Waste Service collects recycling in a converted milk float. This means our local team is visible in the street all day and is able to answer people's questions. Instead of being hidden in a black bag, refuse for landfill is collected in a clear bag as a prompt to citizens to take out that last bit of material for recycling. An opaque blue bag is provided for private or messy waste. This innovation alone led to a 10 percent increase in recycling.
2. **Creating Local Value.** In addition to providing local jobs and training opportunities, the service provides a community dividend from the sales of recycled materials. This goes back to community groups, with everyone in the community able to vote for who should get the money.
3. **Involving People.** Many people in the town are involved in the service. A local farmer lends his equipment to load the paper truck at the depot;

FIGURE 10.13. Cwm Harry workers deliver compost to Cwm Harry's allotments, spread it onto vegetable beds, and reap bountiful results. *Photograph courtesy of Cwm Harry*

growers at the community allotment use Cwm Harry's compost made from food waste; people attend the "bring takes" where you bring the stuff you don't want and can take away stuff you do; and volunteers help distribute bins, sort clothes, and run events.

The town crier helped get the message across during the Waste in Your Face campaign (see figure 10.15). Waste from individual streets was placed in steel cages and called out by the crier: "Oyez, oyez! One tonne of rubbish going to landfill from this street! About 60 percent of this waste could be recycled! There are three homes in this street that don't recycle!"

One group even made a giant puppet out of recycled materials and handed out recycled fabric bags to mark the introduction of the area's new charge on single-use bags.

The service—now nearly two years old—has shown how a community can make a genuine impact. Being given the rights and responsibilities over their own resource management has resulted in astounding reductions in material consumed, down by 24 percent. And of what they do consume three-quarters is recycled, giving a financial return to the community.

Members of the community are now asking Cwm Harry why they can't get big business to play its part. Why, if the community can do this much, can't industry design out the remainder of the waste? That is why we are seeking funding for our next project: a proposal to build a "little museum of bad design."[15]

FIGURE 10.14. Source-separated recyclables stored in Cwm Harry's depot ready for market. In the background are aluminum cans and in the foreground aluminum pie plates. *Photograph courtesy of Cwm Harry*

FIGURE 10.15. *From top:* Cwm Harry's Waste in Your Face campaign involved a tell-all town crier, caged displays of neighborhood waste, and a giant puppet handing out reusable shopping bags. *Photograph courtesy of Cwm Harry*

Wales

Cwm Harry is a nonprofit organization and is running an inspiring program in the community of Presteigne and Norton, which is located about a forty-minute drive from the Welsh-English border near Shrewsbury. The Cwm Harry operation epitomizes a waste program conducted in the community interest as opposed to the corporate or bureaucratic interest. It is a program that has minimized the use of expensive technology and maximized the involvement of local citizens. In addition to the jobs it has created, any profits from the operation have been plowed back into other social programs. The project also collects impressive research data about the residual waste fraction and uses it to improve the system—precisely the kind of feedback mechanism I had in mind when recommending the building of residual separation and research facilities in front of existing landfills. Moreover, Cwm Harry is spreading its model—cooperating with places like Coventry University and Capannori in Italy to create a similar research operation. In the sidebar on Cwm Harry, Katy Anderson, the director of the program, describes some practical details of its operation.

KERB-SORT

Kerb-Sort is another development that has emerged in Wales—a commercial extension of the slow recycling principle promoted by Cwm Harry. Commercially built Kerb-Sort vehicles (see figure 10.17) are designed so that recyclables can be sorted right at the curb.[16] A curbside collection system using this vehicle has recently been introduced in Ballymena, in Northern Ireland.[17]

The Kerb-Sort slow recycling approach has received a nod of approval from the Welsh Assembly Government. In its municipal sector plan in a section titled "Towards Zero Waste: One Wales—One Planet" we find this statement:

> Evidence gathered by the Welsh Assembly Government to date indicates that the best way of delivering the sustainable development outcomes laid down in Towards Zero Waste is for recyclable materials to be sorted at the kerbside. This helps the achievement of high quality source separation of materials and closed loop recycling.[18]

This Welsh government document further explains that closed-loop recycling happens when a material is recycled back into the same product, for example, whereby an aluminium can is used again as an aluminium can and

Contradictions in European Policies

BY JOAN MARC SIMON, COORDINATOR OF ZERO WASTE EUROPE AND GAIA

The European Union is known to have the most advanced environmental policy in the world and has been pioneering waste management for decades. While some places in the European Union (e.g., Italy and Spain) have implemented the most efficient and ambitious source separation and prevention programs, the European Union and its member states have also relied heavily on incineration to burn the residual waste as well as a good part of the recyclable and compostable fractions.

The Waste Framework Directive, a European law (2008/98/EC), established a binding waste hierarchy that gives priority to waste prevention, reuse, and recycling (in that order) over incineration with energy recovery and landfills, and incinerators without energy recovery. However, as nice as this can sound there are three big contradictions with European Union waste policy.

Contradiction Number One

The EU waste policy has deviated too much toward energy policy, giving overwhelming priority to energy generation over energy conservation. Indeed, it is scientifically proven and generally accepted that prevention, reuse, and recycling have a bigger impact on energy savings via the conservation of embedded energy in products than recovering energy by burning them. These savings from embedded energy are three to five times greater than the energy that is recovered when burning the waste. However, EU law doesn't reward the energy savings associated with prevention, reuse, or recycling but does reward generating energy from waste. In the EU directive on renewable energies, burning waste made of biogenic carbon—paper, green waste, kitchen waste, and other nonfossil carbon—is considered to be carbon neutral and hence a source of renewable energy.

In figure 10.16 it can be seen how important biomass burning is for the EU statistics on renewables: 69 percent of EU renewable energy comes from biomass. The "renewable energy" generated by burning waste is of the same order of magnitude as what the European Union generates from wind power.

Consequently, the energy from incineration generated from burning biogenic carbon gets premiums per kilowatt/hour, whereas the energy saved by recycling paper or composting kitchen waste is not rewarded at all. This is clearly a perverse incentive and a contradiction with the European waste hierarchy.

Contradiction Number Two

The second contradiction is once again related to economic incentives. Whereas the European Union continues to promote recycling and green growth, the EU funds designed to help less-developed countries catch up with the most-developed ones have been traditionally used to finance the construction of sanitary landfills

Consumption of Renewable Energy, EU-27, 2010

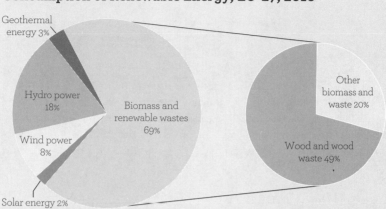

FIGURE 10.16. Renewable energy production in the European Union. *Source: Eurostat (online data code: nrg_1071a, nrg_1072a)*

and incinerators. At the same time, and despite waste directives calling for 50 percent recycling of MSW and specific targets for plastic, glass, paper, and metals, the EU investment in prevention, reuse plants, recycling and sorting facilities, composting plants, anaerobic digesters, and the like is very meager. So while EU money is helping to fund as many as twelve new incinerators for Poland and three in the Czech Republic, very little is going into funding recycling infrastructure. So the rhetoric is going into recycling, but the money is going into incineration.

This contradiction is particularly glaring at a time in which there has been a huge stimulation in the European market for incineration thanks to the reclassification of those incinerators with the highest efficiency from disposal to recovery. The numbers don't match the rhetoric. The European Union already has a capacity to burn 22 percent of its waste, so considering that more than 80 percent of the current MSW in the European Union is either recyclable or compostable one wonders why the European Union wants to continue to build incinerators while claiming to want recycling rates to go up. The European Union can't do both.

Contradiction Number Three

The third contradiction is the opening of a European market for incineration while pretending to handle waste as close as possible to the source. An open market for goods can be positive because it allows resources to go where they are needed at a competitive price; however, an open market for "bads," such as waste, is not a good idea. The fact that the price of the "bad" is not positive but negative (one doesn't buy waste but rather gets money to handle it) means that it provides the wrong incentive for a market to function. What sense does it make to ship waste from Naples, Italy, to be burnt in the Netherlands, as is happening today—especially when most of this waste is recyclable if properly separated?

The existence of this European incineration market can be interpreted as a sign of failure of EU waste policy. On the one hand the European Union has built a huge overcapacity for incineration in countries such as Germany (from 2 to 4 million tons of overcapacity), Sweden, and the Netherlands—and on the other hand it has failed to achieve effective separate collection schemes in most of the European Union. The European Union has an undercapacity as far as waste prevention policies are concerned. Ecodesign directives are still insufficient, and far too few reuse centers, recycling, composting, and anaerobic digestion facilities have been built. The perverse incentives for incineration have been confusing a sound and sustainable EU waste and resource policy.

The Good News

Fortunately, EU waste policies are being taken over by the EU Roadmap to a Resource Efficient Europe, which lays out the steps that the European Union should follow in order to deal with its unsustainable use of resources and talks about "introducing a general ban on waste landfill at the European level and for the phasing-out, by the end of this decade, of incineration of recyclable and compostable waste." As recyclables and compostables make up about 90 percent of the discard stream in the European Union, the EU Parliament has virtually spelled out the end of incineration in Europe.

FIGURE 10.17. A Kerb-Sort vehicle operating in Wales.

The ideas of the roadmap still need to be turned into EU law, but they give clear signs that if the European Union is to be sustainable it needs a zero waste policy in line with what zero waste groups have been advocating since the late 1990s.

The Roadmap to a Resource Efficient Europe also acknowledges that funding should be redirected to composting plants, reuse centers, and similar measures. However, as yet there has been no official consideration given to rewarding the energy savings of prevention, reuse, or recycling. Nor has there been discussion of removing the preference given to energy generation via incineration.

The Zero Waste Network in Europe

It is also encouraging to see a growing zero waste network in Europe that is proving that separate collection rates above 85 percent are possible, that waste can be managed locally and close to the source (providing sustainability and creating local jobs), that no more incinerators are needed, and that it is more efficient to design waste out of the system at the industrial level than trying to perfect ways to dispose of it at the community level. The European Zero Waste Network is trying to to increase the recycling targets in the European Union from the current 50 percent to 75 percent by 2020; introduce mandatory separate collection of organic waste; shift from funding big-scale capital-intensive infrastructure to small-scale labor-intensive "reduce, reuse, recycle, and compost" solutions; reward energy savings more than energy generation; and work upstream by advocating incentives for good product design.[19]

glass from bottles is recycled back into new bottles. This replaces the need for virgin materials in that product (which require a lot of energy to extract from the environment and process), and therefore closed-loop recycling generally provides the greatest environmental benefit.[20]

However, just as in Vancouver and the Toronto region in Ontario, these lofty claims for a zero waste strategy are contradicted by several efforts to build incinerators in Wales.[21, 22] See also the box titled "Contradictions in European Policies" for further insight on how high-flown rhetoric for sustainability in Europe crosses paths with the nonsustainable practice of incineration.

India, Japan, the Philippines, and Taiwan

The fact that India is facing a flood of incinerator proposals[1, 2] should remind us that the people who are most threatened by these thoughtless developments are the millions of people who earn a living picking out valuable materials and objects from landfills. GAIA, working closely with Women in Informal Employment: Globalizing and Organizing (WIEGO), has waged a campaign to give these ragpickers in India and other countries a voice on the world stage. Their goal is to publicize the important role ragpickers play in reducing greenhouse gas emissions, press climate funds to recognize that role, and push for "a new, non-market, climate finance mechanism . . . to support the formalization and expansion of the informal recycling sector."[3]

In the worldwide zero waste campaign we need to support these ragpickers—particularly as decision makers opt to have their waste looked after by multinational corporations rather than by local people. Perhaps the most tragic example of a lost opportunity was in Egypt, where the government in Cairo sabotaged one of the most efficient—albeit rife with personal health dangers for the participants—waste handling systems in the world (see chapter 12). Imagine what could have developed had Cairo supported the ragpickers, improved conditions for them, and formalized a system that provided both income for workers and waste reduction for the city. Instead, they handed over the reins to three European corporations.

Zero waste is not just about saving material resources, it is also about saving people. In addition to helping fight off incinerator proposals, the zero waste movement needs to encourage host countries to provide better working conditions for the ragpickers. They need showers, safety equipment, better accommodation, education for their children, legal access to discarded materials, and profit-sharing arrangements. They also need a seat at the table when waste management plans are being discussed. Most of all they need our respect (see figure 11.1).

The Fifth R

As we expand our discussion of zero waste into the Global South it is time to add the fifth R into our discussion. Most people have heard of the three Rs of reduce, reuse, and recycle; previously I added the fourth R, redesign, to

FIGURE 11.1. Ragpickers rally for dignity in India. *Photograph by Amit Thavaraj, KK PKP/SWaCH*

recognize the vital role of industrial responsibility needed to reach zero waste. Now we must add the fifth R of *respect*: respect for ragpickers, who play a vital role in the Global South.

My ideal for ragpickers would be instead of picking at the landfill itself these people would man the residual screening facilities built in front of the landfills. The costs to government of doing all these things would be miniscule compared to the millions they are prepared to fritter away on incinerators.

Once they have a decent foothold in a clean operation then hopefully these ragpickers can be offered career advancement opportunities by working in reuse and repair operations—as well as the many other activities that can spring from these. Ironically, Daniel Knapp started his career in waste (with nothing in his pocket except a PhD in sociology) by picking out valuable materials from the San Francisco Landfill. After thirty years, his Urban Ore is a thriving business with twenty-seven full-time and well-paid employees.

Neil Tangri of GAIA stresses that there is no single model for incorporating the ragpickers into more equitable and sustainable waste management practices.

> The demands of waste pickers vary widely from place to place as
> does their degree of organization and politicization, so rather than

suggest employing them in an up-front screening facility, I would say that they should be included in the design and planning processes, so that they can integrate as they best see fit. There are some really interesting and different models of how that is happening.

Many of the waste picker groups themselves are quite well organized and powerful. The Movimento Nacional dos Catadores de Materiais Recicláveis (MNCR) in Brazil counts almost a million members and is heavily influential in national policy-making.[4]

India

Tangri provides an inspiring description of the twenty-year history of the ragpickers' union in Pune, India (Kagad Kach Patra Kashtakari Panchayat), in the opening case study of GAIA's recent report on zero waste cases from around the world.[5]

He points out that the ragpickers' organization is working closely with the local municipality to institute door-to-door collection of source-separated waste, in a program known as Solid Waste Collection and Handling (SWaCH). For a small fee, the ragpickers (who are now collective members) pick up organics and dry waste separately. They sort and sell the dry stuff for recycling; the wet waste (mostly food scraps) goes to either local compost pits or one of the small biogas facilities run by the municipality.

Tangri explains how "one of the city's most marginalized and vulnerable populations has become integrated into society" and shows how the current program saves the city an "estimated US$2.8 million per year." He points out that by anaerobically digesting the organics, methane (an important greenhouse gas) emissions are lowered. The program has also produced energy savings and relieved some of the pressure on natural resources such as forests. As SWaCH grows, Tangri predicts that it will mean fewer waste containers on the streets, lower disposal fees for residents, and less waste being burned on the side of the streets.[6]

In the same GAIA report Virali Gokaldas writes about the ragpicker operations in Mumbai. In particular she writes about one nongovernmental organization called Stree Mukti Sanghatana (SMS) that has been training and organizing women ragpickers since 1975. This program has large social ramifications because 85 percent of the ragpicker population is poor, low-caste women. To fight the negative connotation of the local community they have been given a new name—*parisar bhaginis*, or "neighborhood sisters." Since 1998 SMS has taught them the principles of zero waste. They learn "how to sort and handle waste from multi-family dwellings, as well as composting and biogas plant management, gardening, and how to organize as worker cooperatives and negotiate contracts." Gokaldas adds that SMS also helps with "contracting and marketing for individual workers and cooperatives."

Gokaldas explains how some of these neighborhood sisters have learned to operate biogas generation plants, which produce methane and manure from a variety of organic waste materials. One of the facilities she describes is the Nisargruna Biogas Plant. She explains:

> [This] plant was developed to convert on-site organic waste at an individual institution or apartment building into useful methane and high-quality manure (fertilizer) to then be sold back to households or local businesses. It was designed to digest almost any biodegradable waste including kitchen waste, paper, animal dung, bio-sludge, poultry manure, agro-waste, and biomass. The plant design is highly scalable and can be made to handle 1, 2, 3, 4, 5, 10, or 25 metric tons of segregated biodegradable waste. A plant processing one ton of waste requires 10 KWh of electricity per day. Generally, the plant returns 10 percent of processed waste as nitrogen rich compost.[7]

The ragpickers are not the only source of recycling in India. Millions of householders (or their servants) take their recyclables to "recycling shops" and receive a payment for doing so. The size of the payment depends on the type of material delivered.

Kovalam, Kerala

A number of years ago I was invited to give a presentation at a zero waste conference held in the holiday and fishing resort of Kovalam in the state of Kerala. To my surprise and delight I found that this conference had been sponsored by local hotel operators. The reason for their concern was the fact that their beautiful beaches were being littered with plastic water bottles and plastic bags. Needless to say this was neither creating a good impression with tourists nor making the local fishing industry happy. The first suggestion to solve this problem was to burn this waste in a small incinerator. This immediately generated a lot of concern among local citizens and environmental groups. Out of the furor came the suggestion that Kovalam should instigate a zero waste program. This conference I attended was the attempt to put flesh and bones on that idea.

Some of the hotel operators had already demonstrated that they were in favor of good environmental solutions by putting anaerobic digesters (biogas plants) on their property to look after the food debris from their kitchens and restaurants. These were working very well, with the hotels being able to use the methane generated to provide the heat for cooking, and the solid residues on their vegetable gardens: a nice closed-loop system.

Kovalam did indeed embrace a zero waste program and created a number of jobs for local townspeople in the process. They set up several cottage industries to make items out of waste products, including shopping bags out of old newspapers to replace plastic bags. I was recently presented with a product from one of these local craft industries: a small cloth bag to carry a cell phone around the neck. It is colorful, attractive, and very functional. Other craftspeople are making all kinds of products out of coconut shells, which are produced in abundance in the area. In addition the hotels have put in water fountains, allowing their guests to fill and refill their own reusable water bottles.[8]

Japan

The last place I expected to find any discussion of zero waste was Japan. After all, at one time this country had three times more municipal waste incinerators operating (about 1,800) than the rest of the world combined. Tokyo has twenty-three incinerators, and you can see twelve of them when driving between its two airports. However, as you will see from Gary Liss's essay in chapter 22, Japanese industries have been at the forefront of companies declaring a zero waste strategy.

Kamikatsu

Kamikatsu is a little community on one of Japan's southernmost islands. It has not only declared a zero waste goal but has also set up a zero waste academy to monitor its progress. I have visited this community twice—once just before they declared zero waste and once soon afterward. When I last visited the town it had a recycling depot (see figure 11.2) that required people to separate into twenty-eight different categories of discards; it has now pushed that total up to thirty-four categories.[9] All the local shop owners are involved in the program and proudly fly their locally designed zero waste flag outside their stores. They are doing their best to avoid plastic bags. In addition, all homeowners are required to compost in their backyards. Some of them are using fairly sophisticated equipment to do this. I saw backyard composters there that used electrically driven turning devices.

FIGURE 11.2. The recycling center in Kamikatsu, Japan, receives thirty-four different categories of materials. *Photograph courtesy of Zero Waste Academy Japan*

The Philippines

I traveled to the Philippines several times in the 1990s at the invitation of Greenpeace coordinator Von Hernandez, who worked tirelessly with both the grassroots organizers and the political leadership in the Senate to enact a ban on waste incineration. He was eventually successful, and the Philippines became the first country in the world to ban waste incineration. However, the central government has been very slow to invest some of the huge amount of money that would have been frittered away on these costly facilities into infrastructure developments to support aggressive recycling, reusing, and composting programs within the subdivisions or villages (called *barangays*) in its crowded cities. Such investment would have created a huge number of jobs and stimulated many small businesses. Instead, giant landfills have been allowed to swell and fester, not only producing a number of deaths from accidents but also fueling calls for a lift on the incineration ban, which are being stoutly resisted.[10]

Despite this gloomy situation there have been some positive examples of small towns pursuing zero waste. However, these have come about largely because of enlightened local political leadership combined with efforts of the grassroots movement, rather than from pressure from the central government.

One of the organizations that has helped push and develop zero waste programs in the Philippines at both the local and national level is the Mother Earth Foundation cofounded by Sonia Mendoza. See the sidebar on Puerto Princesa to see how local communities are doing on the zero waste front.

Now two cities, Puerto Princesa and Alaminos City in Pangasinan, are engaged in a friendly competition to become the first zero waste city in the Philippines. While vastly different in terms of income, population, and land area, Froilan Grate, president of Mother Earth Fountation, says, "these two cities have one thing in common, aside from being top tourist destinations in the country: they both see zero waste as a solution to their waste problems. Puerto Princesa started earlier than Alaminos. In fact, barangay officials from Alaminos visited Puerto Princesa to learn from their experience. Now, Alaminos appears to be catching up and may be poised to overtake their role model."[11]

Who will get there first? Grate looks at the two cities in more detail and compares their progress. In a personal communication he gave us a number of details.

Puerto Princesa consists of sixty-six barangays and has a population of 210,000 and a land area of 980 square miles. Alaminos has thirty-nine barangays, a population of 80,000, and a land area of 64 square miles. Puerta Princesa has a higher average income than Alaminos.

In this little competition Puerto Princesa began with a clear advantage. First, the city had started much earlier than Alaminos. In fact, they had started to establish model barangays as early as 2007. Second, they had more money available. In total, they had the equivalent of $500,000 to spend on their solid waste management budget.

Alaminos, on the other hand, in early 2010 had only one barangay that had started to construct its own MRF, and only fourteen of the thirty-nine barangays were covered by the city's collection of residual waste. The residents of the remaining barangays were left with the option of dumping mixed waste in their backyard pits or burning it.

Alaminos, however, has the advantage of having a single organizing authority with a clear consensus to pursue a zero waste goal. With some training from GAIA they very quickly finished constructing MRFs for twenty barangays, and more are being constructed. It has less money to spend than Puerto Princesa, but they are spending it wisely. Each barangay is being allotted enough money to allow for the construction of an MRF and the purchase of a collection vehicle. Also, the city has increased its coverage for the collection of the residual waste and aims to serve 100 percent of the barangays very soon while still using the

Puerto Princesa on the Road to Zero Waste

BY FROILAN GRATE, PRESIDENT, MOTHER EARTH FOUNDATION

Mayor Edward Hagedorn of the city of Puerto Princesa has partnered with Mother Earth Foundation (MEF) to help the different sectors in the city, especially the barangays, to embark on the path to zero waste and comply with the provisions of the Ecological Solid Waste Management (ESWM) Act.

The goal of the partnership is to make Puerto Princesa one of the cleanest and greenest cities in the country. Since a sanitary landfill, the first of its kind in the Philippines, had already been constructed prior to the passage of the ESWM Act, one of the goals is to make sure that only residual wastes are put in the landfill, as mandated by law. The project also plans to:

1. establish a material recovery facility (MRF) in every barangay,
2. fully comply with the at-source-segregation provision of ESWM, and
3. achieve a 90 percent diversion rate for the entire city.

After five months, the partnership has shown some signs of success. Already, 90 percent of barangays (from 10 percent prior to the partnership) have at least one MRF. A number of barangays have also removed trash bins along the street to make sure that residents do not throw out mixed wastes and instead bring their segregated wastes to the barangay MRF.

Some barangays have taken a prominent lead in these developments, and these serve as models for others to copy. In Barangay Pag-asa the MRF is being used to compost biodegradable wastes, collect recyclable materials for later sale to junk shops, and act as a central collection point for residual wastes for pickup by the city.

There are also efforts not only to manage waste but also to reduce waste. Chairman Roy Ventura of Barangay Maunlad has instituted steps to be the first barangay to regulate the use of plastic, heeding a challenge issued by the MEF to all barangays as well as to the city council.

This partnership has shown that with the political will and the leadership of the mayor, the decentralization of waste management, the empowerment of barangay leaders, and the involvement and support of all sectors—businesses, schools, and civil society—ESWM is possible now, and zero waste may be more than a distant dream.[12]

same number of trucks and collectors. It is now in the process of looking for technologies to address the handling of these residual wastes.

In early 2009, Puerto Princesa had only six of its sixty-six barangays compliant with the Ecological Solid Waste Management (ESWM) Act. However, by 2012, sixty barangays had established their own MRFs and some *puroks* (a

FIGURE 11.3. An ecoshed (MRF), composting garden, and collection vehicle in an Alaminos barangay.
Photograph by Anne Larracas

FIGURE 11.4 Sonia Mendoza pictured with the Ms. Earth winners. This competition was part of the campaign to ban plastic bags in the Philippines.
Photograph courtesy of Zen Borlongan

smaller subdivision of a barangay) have achieved 90 percent waste diversion. The city now collects the residual waste from sixty of its sixty-six barangays, but it still accepts mixed waste at its landfill, which operates with multiple liners and a leachate treatment plant.

During the Earth Day celebration of 2010, Mayor Hagedorn awarded the model barangays a cash prize—and the six noncompliant barangays were sent notices that they would be sued by the Environmental Legal Assistance Center (ELAC). By 2012, the compliance with the ESWM Act reached 100 percent of the residents.

In conclusion, Puerto Princesa clearly has many advantages, but with Alaminos spending its money very wisely and frugally and with an overall commitment to zero waste and a single organizing authority we might be witnessing a rerun of the hare and the tortoise. As they say, watch this space![13]

Three key features that make the zero waste program visible and tangible to the local population in Puerto Princesa and Alaminos are the locally made ecosheds (or MRFs), the composting gardens for kitchen scraps, and the collection vehicles, which have to be small enough to negotiate the narrow pathways within the barangays. Examples of all three are pictured in figure 11.3.

Banning Bad Packaging

Zero waste advocates in the Philippines have also launched initiatives to ban various forms of bad packaging. The MEF, inspired by Sonia Mendoza (also a member of the Task Force Plastics, EcoWaste Coalition), has been relentless in its efforts to get plastic bags and Styrofoam (polystyrene) banned. Again frustrated at the national level, citizens have turned to local leaders to enact these bans. However, they have had one victory,[14] and such bans are catching on around the world.[15] In figure 11.4 Mendoza is pictured with the winners of the Ms. Earth competition, which was an effort to draw attention to the effort to ban plastic bags in the Philippines.

Taiwan

When I went to Taiwan in the mid-1990s, one of the first buildings I saw as I was being driven around the capital city of Taipei was a huge incinerator. To my amazement the incinerator had a revolving restaurant on its smoke-stack. Such a design, I surmised, suggested that the authorities in Taiwan had overconfidence in this technology. This impression was reinforced the next day when I spoke with the country's environmental protection agency and discovered that they were planning to burn 90 percent of the island's waste.

That would have made this country (population 23 million) the number one burning country in the world.

However, since then I am happy to have found out that the Taiwan Research Institute (a member of GAIA) has led a successful grassroots campaign to partly reverse this situation. This institute summarizes the results of their campaign:

> Between 1997 and 2004, community groups had worked hard to stop the construction of incinerators by the Taiwan government. By 2002, only 19 of the 36 planned incinerators were actually built. At that time, GAIA members pointed out that the government was spending too much money on incinerators while failing miserably to encourage waste prevention and recycling (budget for incineration was around 3.7 billion NTD [US$ 120 million] compared to 0.1 billion NTD [US$ 3.3 million] for prevention and recycling). Moreover, the institute argued that waste production had been decreasing since 1998 and there was not enough for all the 19 incinerators to burn, and thus continuing to build the incinerators would waste money and destroy resources, while producing toxins like dioxin that would harm the environment and people's health. GAIA members urged the government to adopt the zero waste policy. Finally the government adopted the idea but it would not give up the incinerators that have been built. However, the government stopped the construction of 10 big incinerators and the recycling rate has started to surge.[16]

So while Taiwan is far from a model for zero waste, because it is saddled with so many incinerators, it has at least shifted some of its priorities. Like Belgium it has been creative in the way it has instigated a number of programs to stimulate more recycling, composting, and waste prevention. According to an article in the *Taiwan Review*, the recycling rate in Taiwan has surged from 6 percent to 32 percent in the ten years between 1998 and 2008.[17]

Bahrain, Egypt, and Lebanon

Throughout the Middle East, little action has taken place to establish waste management programs that move in a sustainable direction. There is still a heavy reliance on landfilling with occasional efforts by foreign companies to build incinerators to replace them. Bahrain has embraced the zero waste concept but has not taken actual practical steps yet to launch a zero waste program. At the First Bahrain Waste Management Forum and Exhibition Bahrain's Minister for the Environment Rehan Ahmed said that "reduce, reuse and recycle is no longer enough; we need to add respect, return, refuse and re-design."

Ahmed encouraged Bahrain to press for producer responsibility, listen to the students and other young people who were calling for a zero waste program, and create a program that treats the disease (the sources of waste) rather than the symptom (the waste itself). He upheld zero waste as the key to sustainability and advocated a "total ban on all types of waste burning, gasification, plasma arc, pyrolysis, and other destructive technologies." He also urged industries and the community to become zero waste leaders at work, home, and elsewhere—engaging their entire company or community and educating them on zero waste programs.[1]

Signs like this that the zero waste concept is being increasingly understood and embraced in the region are heartening, but there is much work to be done to turn rhetoric into reality.

Egypt

Cairo used to have one of the most efficient waste handling systems in the world, and it didn't cost the government a penny. For nearly eighty years the Coptic Christian community has hosted the Zabbaleen (literally "garbage people") who have built their way of life around collecting waste door-to-door with trucks or donkey-pull carts, separating out materials for recycling—often converting them into secondary products using their own acquired equipment—as well as feeding the organic waste to their pigs. Overall their diversion rate was around 80 percent. Hotels in a Red Sea resort had even called upon some of the members of the community to help solve the waste crisis that was threatening tourism in the area. Piles of plastic bottles and other debris were cluttering their beautiful beaches.

Sabotage of the Zabbaleen System

The Zabbaleen's informal system of managing waste has been sabotaged in two ways. First, the Cairo government brought in three European companies (two Spanish and one Italian) to very lucratively haul away Cairo's waste and landfill it. As a result the city's recycling rate plunged to 20 percent. Second, the national government, in a fit of religious persecution, ordered the slaughter of all their pigs (swine flu was the pretext, but there were no cases). That axed their organics handling, and the diversion rate fell dramatically. Several films have been made about the Zabbaleen's predicament.[2-4]

There is no denying that the conditions in which the Zabbaleen collected waste were horrific (piles of waste built up in the streets, and the stench from the pigs was horrendous), and the relationship between the laborers and some of the small recycling business owners within the community was quite exploitative. There was also the whole issue of child labor. However, had the government worked to address these problems, in concert with organizations offering support on the social front, they could have taken a step in a more positive direction.

Despite these dramatic setbacks the Zabbaleen continue to do what they can to eke out an existence from waste. Recently, Gigie Cruz of GAIA visited the Zabbaleen in Mokattam, Cairo, and documented the recycling and reuse operations still ongoing in the community, albeit at a reduced rate. Nearly all the residents of Mokattam are involved in the waste trade either as collectors, middlemen, or small-scale recyclers within the community. "Being involved in the waste collection business," she notes, "there has been a standing public misconception of the Zabbaleen being dirty and undesirable. Initiatives are being undertaken to empower and improve their conditions by giving them proper education through the Recycling School established with the support of organizations like the Association for the Protection of the Environment and Spirit of Youth for Environmental Services (SOY)."[5]

Cruz has shared with us a large number of photographs she took in the community, and these are shown below along with a few from Laila Iskandar, a social activist who has helped the community in many ways (see figures 12.1 to 12.4).

When one looks at the photographs—especially the aerial shots—of this community, it is easy to get the impression that one is looking at squalid chaos. However, below the surface and inside the rooms of the houses one finds a very industrious and highly organized community that probably knows more about what is in the domestic waste stream and the uses that can be found for its many components than waste managers in any other community on earth.

FIGURE 12.1. This aerial shot illustrates how embedded waste collection and processing is in the Zabbaleen community. Every square inch seems to be devoted to some aspect of resource recovery. *Photograph by Gigie Cruz, GAIA*

FIGURE 12.2. A close-up of some of the Zabbaleen waste operations. *Photograph by Gigie Cruz, GAIA*

FIGURE 12.3. Waste collection in the Zabbaleen community can involve heavy manual labor. *Photograph by Gigie Cruz, GAIA*

Figures 12.5 and 12.6 show schematic summaries of some of the many operations undertaken within the community.

Figures 12.7 and 12.8 show how meticulous the sorting operations for plastics and other items can be. In particular hard plastics are sorted by color and specific use (e.g., shampoo bottles) prior to pelletization. Such specific sorting guarantees a high-quality product for resale after granulation or pelletization because it is of uniform content and color.

FIGURE 12.4. Not only do the workers separate and upgrade the discarded materials, but many families are also involved in processing the materials into new products. *Photograph by Gigie Cruz, GAIA*

FIGURE 12.5. Some of the many sorting and processing operations going on in the Zabbaleen community. *Image courtesy of Laila Iskandar, CID Consulting*

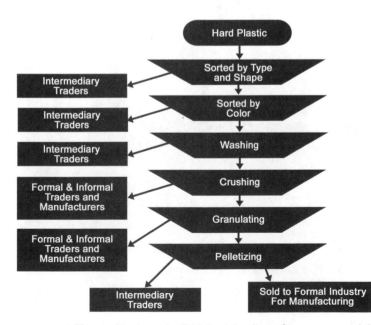

FIGURE 12.6. This graphic shows the Zabbaleen's handling of just one material: hard plastics. *Image courtesy of Laila Iskandar, CID Consulting*

FIGURE 12.7. Familiy members sort plastic, by color, into sacks for further processing. *Photograph by Laila Iskandar, CID Consulting*

FIGURE 12.8. Plastics of a single color are shredded together for reuse. *Photograph by Laila Iskandar, CID Consulting*

Instead of being sold to external processors and manufacturers, some of the plastic products are used in fabrication processes within the community itself. Figure 12.9 shows a process in which plastic coat hangers are being made.

FIGURE 12.9. A press used to make coat hangers out of recycled plastic. *Photograph by Laila Iskandar, CID Consulting*

Figures 12.10 to 12.14 show a whole series of cottage industries making many other products from discarded and separated materials—from paper products to textiles. These operations are far more sophisticated than outward appearances might suggest. The last time I saw paper being made in the manner shown in figures 12.10 and 12.12, it was in the museum in Fabriano, Italy—the home of papermaking in Europe.

While the government has managed to drastically reduce the Zabbaleen's use of pigs to deal with discarded organic waste, some members of the community have turned to goats and sheep to perform the same function (see figure 12.15).

The Zabbaleen have a lot to offer other communities wondering how to add value to separated materials by making new items in the local community. Hopefully, the new government will work more positively and creatively with them in the future and use their willingness to work with waste (or rather, discarded materials) by formally including their representatives in their municipal programs as some South American and Indian communities have done (see chapters 11 and 13). In this way the enormous amount of money being spent on waste could go into improving the lives and working conditions of thousands of people and their families instead of ending up in the coffers of multinational waste companies.

FIGURE 12.10. Two women involved in making paper from recycled fabrics and other fibers. *Photograph by Gigie Cruz, GAIA*

FIGURE 12.11. Papier-mâché wall decorations are crafted from recovered paper products. *Photograph by Gigie Cruz, GAIA*

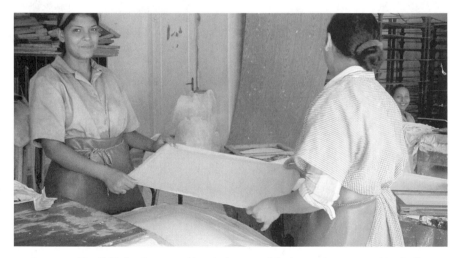

FIGURE 12.12. The Zabbaleen's papermaking rivals some of the best artisan papermaking in Europe. *Photograph by Laila Iskandar, CID Consulting*

FIGURE 12.13. Young women separate rags in preparation for weaving. *Photograph by Laila Iskandar, CID Consulting*

FIGURE 12.14. A young girl weaving discarded fabrics into a colorful textile. *Photograph by Laila Iskandar, CID Consulting*

FIGURE 12.15. After their pigs were slaughtered, some in the Zabbaleen community used goats to deal with the discarded organic fraction. *Photograph by Gigie Cruz, GAIA*

Lebanon

In most of the rest of the Middle East there seems to be still a major reliance on landfills. One of the most appalling of these is operating in the south of Lebanon and is featured in the full-length documentary *Trashed*.[6]

This Lebanese landfill is located right on the edge of the Mediterranean Sea. Every so often huge chunks of the landfill are washed away into the sea (see figure 12.16) and float off to contaminate beaches in other countries bordering the Mediterranean. In addition to the chunks of debris that float out to sea, the leachate that oozes out of the bottom of any landfill is also washing out to sea.

There has been interest in Lebanon for reform. In 2011, Joan Marc Simon and I were invited to speak at a zero waste conference in Beirut organized by local activists to present an alternative to the government's proposal to build an incinerator. The response to the zero waste message from the largely university audience was very positive, but I have yet to hear if it has led to any initiatives being taken by Beirut or any other Lebanese community. Again, there is a world of difference between enthusiasm for zero waste and actually taking concrete action.

FIGURE 12.16. Jeremy Irons, in the documentary *Trashed*, in a landfill on the edge of the Mediterranean Sea in south Lebanon. Leachate and debris from the landfill is continually washing into the sea and contaminating the beaches of many other countries. *Photograph by Laurence Richards, copyright Blenheim Films, all rights reserved*

South America

Nowhere is the zero waste movement's combined quest to save both resources and people more apparent than in the efforts to secure the livelihoods—as well as gaining respect for—the millions of ragpickers in cities in Latin America and elsewhere in the Global South. Several countries in Latin America are doing their best to win recognition for ragpickers and are helping them to organize into cooperatives so that they can be included in local government waste management plans.

The Brazilian researcher Jutta Gutberlet has written movingly about the importance of these ragpickers in her excellent book *Recovering Resources, Recycling Citizenship: Urban Poverty Reduction in Latin America*.[1] When she writes, "Social economy brings social justice issues and values, such as co-operation, redistribution and reciprocity, into the economy,"[2] it reminds me of the subtitle of the late E. F. Schumacher's landmark book *Small Is Beautiful: Economics as if People Mattered*.[3]

Gutberlet stresses again and again that ragpickers are the "key players in the struggle for social justice."[4] She encourages practices that not only view waste recovery as "the necessary end of the product life-cycle" but looks at them as a tool "that implements inclusive waste management as a contribution to social justice." She also advocates helping ragpickers organize "in associations, cooperatives or other forms of community organizations."[5]

In a recent article Gutberlet describes two case studies in Brazil (Diadema and Londrina) where efforts have been made to include ragpickers in the formal solution of the waste and resource problem. She explains the benefits of this "inclusive solid waste management" as

- tackling socioeconomic vulnerability,
- reducing waste management costs,
- promoting greater resource efficiency,
- requiring an intersecretarial and interdisciplinary urban planning and development approach,
- building social cohesion and fostering community, and
- increasing urban resiliency.

Despite the success of the two cases she highlights she indicates that "there are still hurdles that need to be overcome":

- the lack of government/business support, given the prevailing growth-oriented economic development model;
- the need to integrate local recycling initiatives into official waste management programs, in order to play the leading role in waste reduction;
- and the lack of dialogue and integration among the stakeholders involved, which should include consumers, producers, government, recycling initiatives, and the recycling industry.[6]

In this chapter we will look at one example of inclusive solid waste management, in Brazil, but there are two other examples I would cite as well—the first in Buenos Aires, which has a zero waste policy,[7] and the second in La Pintana in Chile.[8]

FIGURE 13.1. A huge rally in Brazil calling for respect for ragpickers (*catadores*, or "collectors of recyclables"). *Photograph by Sonia Dias*

While the diversion rates in Brazil, Buenos Aires, and La Pintana are not high, the efforts described show very important steps in involving and organizing ragpickers in formally solving waste problems. Other countries in the Global South would do well to copy these more humane approaches to harness the willingness of ragpickers to do this often dangerous and little-respected work. We need to improve their working conditions *and* we need governments to give recognition to the importance of their efforts in the move toward zero waste and sustainability worldwide. Figure 13.1 shows a huge rally organized by ragpickers in Brazil.

Brazil

Sonia Dias, a waste sector specialist working in the Urban Policies Program of Women in Informal Employment: Globalizing and Organizing (WIEGO), has written extensively on social issues in Brazil, including efforts to secure recognition of the informal waste management sector.[9-16] Below, in the box on Belo Horizonte, she reports on efforts underway on recycling initiatives in Brazil. She stresses that these notes on the Belo Horizonte case study are intended to be demonstrations of new ideas and not to serve as a model to be applied in its entirety elsewhere since the physical elements of solid waste systems as well as the policy, institutional, and governance frameworks differ from one place to the next.

The level of organization of informal workers also varies. However, it is hoped that the example of Belo Horizonte will give ragpickers, their organizational leaders, and policy makers in other locations ideas that could be adapted to their local realities.

Colombia

Ragpickers have also organized themselves very effectively in Colombia. In 2011, they successfully fought a $2.5 billion public bid that would have taken the role of recycling from the *recicladores* and handed it over to private companies for a ten-year period.

In Bogotá, the ragpickers make little money because most of the earnings go to middlemen who buy the scraps from them and sell it on to industry. So they are forming cooperatives, renting small industrial spaces and vehicles, and investing in machinery (like vats for Tetra Pak recycling or a chipper for PET), which allow them to sell directly to the factories—significantly upping their earnings. Faced with the privatization of the waste management system, they have been going to court to defend their right to recycle.[17, 18]

Recycling in Belo Horizonte, Brazil: An Overview of Inclusive Programming

BY SONIA DIAS

Brazil is one of the world's most progressive countries in integrating ragpickers into solid waste management systems, and Belo Horizonte has led the way. The capital city of Minas Gerais State in southeastern Brazil, Belo Horizonte has a long tradition of strong municipal planning that dates back to its development in the nineteenth century as Brazil's first "planned" city. Waste management has been a municipal priority and concern in the city since 1900.

With a population of approximately 2.5 million people, the city generates 1.3 million tons of municipal waste a year, and it has been a pioneer in managing solid waste in Brazil. Socioenvironmental concerns, such as improving the existing systems and generating income for the poor, encouraged improvements in solid waste management (SWM) systems in the city. This led to the adoption of an integrated SWM model in 1993, with a focus on promoting segregation at the sources in order to minimize the harmful environmental impact caused by the waste itself and maximize the social and economic benefits for the city. The new integrated system brought wide-reaching improvements, including:

- enhancements to the operations at the existing landfill;
- selective collection and a recycling program for civil construction waste;
- composting of organics;
- environmental education;
- improving working conditions for formal workers (sweepers and collectors); and
- integrating informal workers into the formal SWM system.

At the municipal, state, and national levels in Brazil, there has been a commitment to social inclusion of residents. The integration of the informal recycling sector in SWM systems is a good example of this. Since 2001, for example, Brazil has included *catador de material reciclável* (meaning "a collector of recyclables," informally called a *catadore*) as a profession in the Brazilian Occupation Classification (CBO). The CBO describes an individual with this job as "someone who might collect recyclables in streets or at disposal sites, work as a sorter and/or other recycling related activities either in cooperatives or junk shops." Years earlier, Belo Horizonte had introduced legislation that made recycling, social inclusion, job creation, and income generation the four main pillars of SWM.

The main feature of these recycling programs is the integration of two categories of informal workers: the ragpickers (*catadores*) of recyclables, and the informal collectors of debris (*carroceiros*).

In Belo Horizonte 95 percent of the households have waste collection, 2,685 catadores are working in recycling, and there are eight recycling cooperatives with 401 affiliate organizations. Of the catadores, 15 percent are organized into associations and cooperatives.

SONIA DIAS

The Public Cleansing Authority (SLU) has the mandate for providing all solid waste management services (sweeping, collection, disposal, treatment, and transformation of garbage) for the entire city of Belo Horizonte, including the sale of its services, products, and by-products. The SLU defines domestic waste as waste generated by households, as well as waste with similar characteristics generated by commercial establishments, institutions, and industry. Recycling activities focus on domestic solid waste and construction waste.

FIGURE 13.2. Dias (center, back row) with some of the *catadores* for whom she works to gain governmental and public respect. *Photograph by the Public Cleansing Authority (SLU)*

Organizations of informal workers (catadores and carroceiros), together with representatives of the municipality and other organizations of civil society, have joined one stakeholder forum: the Municipal Waste and Citizenship Forum.

A critical source of success of these programs is the high level of organization and social mobilization of the ragpickers and their supporting nongovernmental organizations.

Recycling of Materials from Domestic Waste

The municipality recovers nonorganic recyclable materials from the domestic solid waste stream through three main channels: a drop-off system, a curbside separated collection of recyclables in residential areas, and door-to-door collection from nonresidential facilities.

All collected materials (mainly plastic, paper, and metals) are brought to warehouses of ragpickers' cooperatives. Here, the materials are processed before moving up the recycling chain. All cooperatives have scales, personal protection equipment, and big bags. Some have shredders and forklift trucks. Materials are sold to industry in Belo Horizonte or within the state of Minas Gerais. The cooperatives receive all the money from sales, which is then shared among the associates.

This official integrated recycling system combines the work of the formal sector and the semiformal sector (cooperatives) through several channels.

Door-to-Door Collection

Two of the eight cooperatives of ragpickers involved with the municipal recycling scheme collect recyclable materials from commercial establishments and offices, especially in downtown Belo Horizonte, using pushcarts. One of these—the Association of Paper, Carton, and Recyclable Material Pickers (ASMARE)—has its own carpentry workshop where carts are made and repaired. Recyclables are collected from larger generators such as industries and public offices using vehicles owned by the cooperatives. The collected material is taken to warehouses for further handling. In 2008, there were 5,100 tons collected through the cooperatives, which made up 52 percent of all collected recyclables.

The Drop-off System

There are 150 delivery sites scattered throughout the city—locally known as Local de Entrega Voluntária (LEV)—where people can bring recyclables and place them in different containers for plastic, paper, metals, and glass. The SLU staff empty about 450 containers each week, and the materials are transported to warehouses for further handling. One challenge has been that many of the recycling containers are damaged, usually by nonorganized ragpickers searching for recyclables.

Curbside Collection

When the SLU started this system in 2003. it served 80,000 people; it now collects from over 148,000 people. The collected materials are taken to warehouses run by the cooperatives of semiformal ragpickers, where they are processed and sold to industry.

Collected recyclable waste materials are brought to the recycling warehouses of one of the eight ragpickers' cooperatives, where the materials are sorted, baled, shredded, packaged, and stored. Most recycling warehouses were adapted, not purpose-built for these activities. As a result, there have been challenges with handling and flow of materials. However, new warehouses are being built that are specially designed for recycling activities and have mechanical discharge and sorting systems. These will be better suited for ragpickers' needs and improve the efficiency of the system

Recycling Construction and Demolition Waste

The city's by-laws state that business owners are responsible for disposing of construction and demolition waste; both business owners and informal waste workers often dumped it illegally until the city, through the SLU, implemented the Environmental Recovery and Recycling of Civil Construction Wastes Program in 1993.

The program includes carroceiros, who use horse-drawn carts to transport collected construction debris. The municipality provided them awareness-raising on the negative impact of illegal dumping, and also organized and registered them. Now citizens have access to this municipal registry and can call upon carroceiros' services for the removal of construction waste, tree pruning remains, or even old furniture. Households pay a fee per trip directly to the carroceiros to transport this special waste to designated reception units.

The city has also provided twenty-nine decentralized depots where the carroceiros can take small volumes of debris, which is then moved to municipal recycling units by city trucks. The city issues the carroceiros an identity card and allows them to register, license, brand, and vaccinate their horses free of charge. Through the city's partnership with the Veterinary School of the Federal University of Minas Gerais, the carroceiros receive health care assistance for their horses. The university also does research to improve the strength and ability of the horse breeds.

The SLU, meanwhile, is responsible for checking that all the material collected and transported by the carroceiros meets the regulations. There are three civil construction recycling units that process construction waste taken from the decentralized reception units, from the public-sector construction activities, and also from large private construction companies that have "separation at the source" programs at their construction and demolition sites. In 2008, the three processing units together received and processed 132,934 tons of construction waste, which accounted for 15 percent of this waste stream and 6.5 percent of all processed waste in Belo Horizonte. These recycling units process bricks and other construction materials out of the debris. The remaining construction waste goes to the sanitary landfill operated by the municipality of Belo Horizonte.

Composting

Tree pruning waste is transported to a small composting facility (capacity of 20 tons/day) at the municipal waste treatment center, where it is coprocessed using a windrow composting technology that also uses organic waste collected from supermarkets and open vegetable markets. A strong monitoring and quality control process is in place to assure a good product suitable for agricultural use. In 2008, 2,300 tons were processed—1,400 tons from markets and 900 tons from tree trimmings. The nearly 900 tons of compost produced in 2012 were used in the landscaping project of the landfill, as well as by the municipal nursery and public gardens.

Behind the City's Success

Like many other cities, Belo Horizonte's methods of dealing with solid waste are designed mainly to ensure public health and focus on waste collection and safe disposal. In 1993—long before the 2010 approval of Brazil's National Solid Waste Policy, which ensures the rights of informal recyclers—Belo Horizonte was already using a progressive approach to SWM. Its Organic Law and other related legislation included recycling, social inclusion, job creation, and income generation as the four main pillars of SWM.

Why has Belo Horizonte been so successful in including informal workers in the waste management processes of the city? One reason is that institutionally SWM has a well-established and all-inclusive position within the municipal hierarchy. The SLU has been operating for more than three decades at a high level of independence. Another reason is the high level of organization and social mobilization of the ragpickers and their supporting NGOs. The example set by ASMARE, the first association formed in Belo Horizonte in 1990, inspired other groups of catadores and carroceiros to be formed in the city. The catadores are able to voice their demands and to form strategic alliances. In addition, the two existing Waste and Citizenship Forums (the Minas Gerais State Forum and the Municipal Forum of Belo Horizonte) provide a platform where interests and concerns from civil society actors and public officials can be brought for public debate.

The contribution of informal workers to building strong and sustainable solid waste systems, the reduction of carbon emissions, and a cleaner and healthier environment for everyone has guaranteed a strong and continuing commitment to social inclusion in Belo Horizonte.[19]

PART THREE

Different Perspectives on Zero Waste

All leaders in the zero waste movement want to see cities and towns across the globe become more sustainable. But many have different perspectives on the best way to achieve zero waste goals or are involved in different areas of the movement.

In the next ten chapters, we will hear from ten pioneers of zero waste in North America. They write about their involvement with, and different perspectives on, the zero waste movement. In North America, Bill Sheehan and Helen Spiegelman are to the front-end of the zero waste movement (industrial responsibility and opportunity) what Neil Seldman, Eric Lombardi, Daniel Knapp, Mary Lou Van Deventer, and Buddy Boyd are to the back-end of the movement (community responsibility and opportunity). Gary Liss is every-where—working at the front-end, at the back-end, and in the middle. Jeffrey Morris provides a very clear economic analysis in favor of zero waste, and Richard Anthony explains the efforts to get the zero waste message clearly defined and spread around the world.

In chapter 24 I provide a summary if not quite a synthesis of all ten views and attempt to bridge one significant divide between those who would seek to give the responsibility for all discarded materials (except the organic fraction) to industry (via EPR) and those who would like to ensure that the community continues to use some of these discarded resources to achieve sustainable and local economic development.

Zero Waste and the Local Economy

By Neil Seldman

Neil Seldman is cofounder and president of the Institute for Local Self-Reliance (ILSR),[1] which has offices in Washington, DC and Minneapolis, Minnesota. Seldman was a manufacturer in New York City and lecturer in political science at George Washington University prior to his career at ILSR. He specializes in business start-up and expansion and policies that reinforce local economic development.

NEIL SELDMAN

In this chapter he provides a summary of ILSR's involvement in waste management solutions in the 1970s, which ultimately grew into today's zero waste movement.

The Beginnings of the Institute for Local Self-Reliance (ILSR)

The focused interest in recycling and economic development began with the founding of the Institute for Local Self-Reliance in 1974 in the Adams Morgan community of Washington, DC. ILSR's mission was to make this community of twenty-five thousand residents self-reliant. It developed programs for waste utilization, urban food production, and decentralized energy systems.

By 1980, ILSR had become a national organization as its mission was expanded to other cities and counties in the United States. It focused on the transition from drop-off center recycling, which reemerged in the late 1960s, to curbside collection. In 1974 there were two curbside programs in the United States, by the 1980s there were hundreds, and by the 1990s there were eight thousand. By the 2000s, eight thousand cities and counties had curbside collection of yard debris for mulching and composting.[2-7]

At the same time ILSR worked with citizens in local groups and businesses in thirty cities to stop the building of garbage incinerators. This involved producing how-to manuals for action at the local level, where most of the solid waste decisions are made in the United States. Between 1985 and 1995, largely through grassroots action, over three hundred incinerators (out of about four hundred) were defeated.[8] No garbage incinerator has been built in the United States since 1995, although some older ones have been retrofitted and expanded. More about this historic grassroots effort can be found in chapter 3.

Returning to ILSR's Number One Mission

With the demise of incineration in the 1990s, ILSR was able to return to its focus on recycling and economic development policy and project implementation. Recycling is a primary national strategy for

- greenhouse gas reduction,
- pollution reduction,
- decentralization of the national economy through local production, and
- reduced expenditures for solid waste management by local governments and private businesses.

ILSR has worked on waste policy with major cities and counties throughout the United States, including Atlanta; Austin; Bridgeport, Connecticut; Cleveland; Los Angeles; Philadelphia; Washington, DC; and King County in Washington State.

It has also helped to start or expand businesses in composting, building deconstruction, small-scale paper production, glass and plastics manufacturing, and reuse and repair centers. These areas of economic activity have grown from eighty-six thousand workers in 1970 to over one million today. If the current rate of recycling in the United States of 40 percent (this figure includes construction and demolition debris) is doubled, we can expect at least another one million jobs. For each direct recycling job, ILSR estimates one indirect job is created in the economy. The biggest jobs growth in the past five years has been in electronic scrap, C&D debris, and food discard recovery. In 2002, for example, there were six thousand workers in the electronic scrap sector. Today there are thirty-five thousand.

As the amount of materials available for recycling and composting have increased, so have the industries that use these materials. New companies emerge weekly to process and add value to recovered materials. Today the recycling industry employs more workers than the US auto industry.[9]

Meanwhile, new rules for procurement, for banning materials from landfills, for extended producer responsibility (EPR), for direct economic incentives to households, and for minimum content laws are economic drivers coming from the local level and impacting practice at the national and international levels.

Incineration Again!

Just when we, and others, had thought we had seen the end of incineration in the United States, the country is being confronted again with a whole new wave of proposals of both conventional mass burn and biomass incinerators, as well as gasification, pyrolysis, and plasma arc facilities.[10] Neither economically nor from the perspective of sustainability can such burn facilities compete on a level playing field with intense recycling. However, the promoters are seeking massive subsidies from the federal and state governments in the name of "green" energy. Today ILSR is again working with groups fighting incinerators. In 2011, one large incinerator project has begun construction in Palm Beach, Florida. Citizens and environmental activists opposed this plant, but lost 3–2 at the county commission.[11]

If there is any consolation in the diversion of ILSR's efforts into this time-wasting exercise (surely we shouldn't have to persuade people in the twenty-first century that it is not sustainable to burn finite resources?) it is that it gives us the opportunity to introduce communities confronted with these incinerator proposals to the zero waste strategy and the opportunities it provides for job creation and small business development at the local level. However, we have to do so in the teeth of the well-funded public relations machinery of the incinerator industry that has shown a remarkable ability in the past to co-opt the language of recycling—and now is attempting to do the same with the concept of zero waste.

Language and History

The battle over language has been at the heart of the development of the term *zero waste*. In the early 1980s Charles Gunnerson of the World Bank introduced the phrase *integrated solid waste management* during a research study on worldwide recycling practices.[12] The term initially meant integration of solid waste management into the social, economic, and environmental needs of the community. Soon, the landfill and incineration industry co-opted the term to mean a solid waste management system that was "integrated" by having a landfill, incinerator, and recycling in the mix. Recyclers countered the ideological and self-serving use of this concept. Urban Ore founder Daniel Knapp and Mary Lou Van Deventer called for "total recycling" as an alternative to integrated resource recovery. To grassroots recyclers, integrated solid waste management as defined by the landfill and incineration industry was just a small step up from the "burn and bury" attitude of the post–World War II era. Massive amounts of valuable raw materials were still being destroyed.

By 1995, the contest for the proper interpretation of sustainable materials management of the discard stream took new directions. ILSR, the California Resource Recovery Association (CRRA), and Sierra Club members formed the GrassRoots Recycling Network (GRRN) in a meeting held in Athens, Georgia.[13] The immediate goal was to differentiate grassroots organizations from the National Recycling Coalition (NRC), which had become a corporate recycling organization. (For example, there was no discussion of bottle bills or incinerators at the NRC's meetings—and technical committees were purged of individuals who wanted to increase recycling and composting.)

In 1995, Daniel Knapp returned from a trip to Australia bringing news to GRRN of a new law pending passage in Australia's capital city of Canberra. The No Waste by 2010 law was eventually passed in 1996 (see chapter 5).[14] Meanwhile this new zero waste message spread rapidly via the Internet around the US recycling movement, particularly in California. In 1997, zero waste was the theme of the important annual conference of the CRRA—one of the largest nonprofit recycling organizations in the world. Zero waste has been included in every CRRA annual conference since then.

In 1999, Bill Sheehan, who was then directing GRRN, was able to persuade Paul Connett to attend the 1999 CRRA conference held in San Francisco and interview the leading zero waste theorists and practitioners of that time for what became the groundbreaking video *Zero Waste: Idealistic Dream or Realistic Goal?*[15]

For me, the period between 1995 and 2000 was a most exciting time. Planned incinerators had fallen like dominoes between 1985 and 1995. Meanwhile, new data was coming out on energy and environmental savings from recycling, composting, and reuse. All of this was reframed into the zero waste message. It was with a huge sigh of relief that we at ILSR were able to refocus on our key mission—building local economies through the best use of local capital, labor, and raw materials. We could again focus on creating jobs and new enterprises from discarded materials.

Zero Waste Today

Today the zero waste concept has become conventional wisdom among environmental groups, local government officials, and significant segments of the business community. However, conceptual and ideological issues remain, as the big waste corporations are attempting to co-opt and corrupt the concept by celebrating "zero waste to landfill," as they send waste to incinerators!

In order to keep the language accurate and maintain the integrity of this movement, the Zero Waste International Alliance (ZWIA) was formed with membership from every continent. The planning group of ZWIA adopted a definition of zero waste in 2004 that rejected both landfilling and incineration and called not only for all discarded materials to be reused—and for products to be designed to be reused—but also for people to adopt sustainable lifestyles and practices (see chapter 1 for the ZWIA definition).[16]

Clearly, there is no place in a zero waste strategy for either megalandfills or incinerators no matter how hard spokespersons for this industry attempt to torture the language to allow them a place at the table. Zero waste is a stepping-stone to sustainability, while incinerators and megalandfills take us in the opposite direction.

Today zero waste activists, or zero wasters, see three important aspects of zero waste—inspirational, practical, and peaceful.

Inspirational. Zero waste inspires us to move in the right direction even as we know we cannot reach the ultimate goal of zero. But we can get very close (95 percent diversion rate or better).

Practical. Reuse and repair operations (like The Reuse People of America, Oakland, California; Saint Vincent De Paul, Eugene, Oregon; Urban Ore, Berkeley, California; ReSOURCE, Burlington, Vermont; Second Chance, Baltimore, Maryland; and many others) are creating thousands of jobs in our cities, grossing millions of dollars, and helping people fight poverty every day. Their discard rate is typically less than 2 percent.

Peaceful. The world has a growing population but shrinking resources. The low-lying fruit of virgin resources has been exhausted by two hundred years of industrial extraction. Thus, we are drilling for oil one mile beneath the ocean and fracking for natural gas in places where huge amounts of underground water could be contaminated. The results are higher costs and devastating contamination of water, air, and soil. Zero waste aims to extend world resources for all countries both in the present and in the future. Hence, it is a critical component of the pathway to international harmony.

With so much at stake it would be a tragedy if communities were sucked again into the hugely uneconomic and unsustainable practice of incineration. We do not have another twenty-five years to waste.

Waste Isn't Waste Until It's Wasted

By Daniel Knapp and Mary Lou Van Deventer

Daniel Knapp is a recycling company executive, entrepreneur, applied social scientist, and social engineer. He is the founder of Urban Ore, Inc., located in Berkeley, California, and has been its chief executive from 1980 to the present. In 2006, Urban Ore was included in a list of the "110 Best Things to Do in San Francisco" and named "Best Place to Recycle" by SF Magazine. *Knapp graduated Phi Beta Kappa from the University of Oregon and holds a PhD in sociology.*

I first met Knapp when he came to Canton, New York, in 1986 to help us in our battle to stop the building of a trash incinerator in St. Lawrence County. I have visited Urban Ore on several occasions both at its former and present locations and can easily understand why the Berkeley City Council stated that "Urban Ore is one of Berkeley's treasures." In addition to providing a superb busi-ness model for protecting the planet, both

MARY LOU VAN DEVENTER AND DANIEL KNAPP

Knapp and his wife, Mary Lou Van Deventer, have performed yeoman service in protecting the language used in the world that deals with discarded materials.

Deventer was teaching high school in Michigan when she had an environmental epiphany and life-changing vision. She migrated to California and became an editor for Friends of the Earth, where she had the opportunity to write the chapter on resources for the policy book Progress as if Survival Mattered. *Later she was a policy analyst and editor with the California Office of Appropriate Technology, where she met Dan, who had recently spoken to the National Science Foundation about salvaging for reuse at a landfill. She left the government to work on the staff of the Sierra Club's* Sierra Magazine, *where she became editorial manager. Since 1982 she has worked with Urban Ore, where currently she is the operations manager.*

Elsewhere in this text (see chapters 2 and 6) there has been a description of the operations of Urban Ore, a reuse and recycling operation located in Berkeley, California. Here we would like to ruminate on our experiences over the last thirty years plus, especially as they pertain to the use and the misuse of language in the "waste business."

Resources, Not Waste

Our company, Urban Ore, doesn't deal with waste; we deal with resources. We are open seven days a week, ten-and-a-half hours per day, with people streaming in and out, and loads delivered and taken away. Moreover, here in California, Urban Ore swims in a sea of resource-conserving businesses with names like American Soil and Stone Products, Community Conservation Center, Standard Metal Products, Tri-CED, and Raisch Products.

The name Urban Ore came to Dan after he left his university teaching career and became a landfill scavenger at Berkeley's shoreline dump.

Beautiful Setting, Ugly Reality

The view from Urban Ore's claim on the ninety-acre landfill was inspiring: straight west was the Golden Gate Bridge, to the right the Marin Headlands, to the left San Francisco and the Bay Bridge. Behind to the east was Berkeley, with the University of California's Campanile gleaming whitely—a memorial to the academic career Dan had left behind. A short hike from the garbage tipping area brought Dan to a stone dyke. At its sloping base lapped the green waters of San Francisco Bay; fifty feet or more of garbage was directly underfoot. Dry weather brought dust; wet weather mud. Winds blew almost all the time, stirring things up. Rain or shine, the ground trembled when the compactors rumbled by. Swooping, soaring clouds of seagulls circled overhead and every so often landed all around us.

Resource Competition and the "Lowerarchy"

At their start-up around Earth Day 1970, recycling enterprises didn't look like much. Some waste managers comforted themselves by believing that there was nothing valuable in the waste stream, because if it was valuable it wouldn't be there. (For example, the head of the solid waste transfer stations and landfills in Lane County, Oregon, used this argument during Dan's first meeting with him as codirector of the new Office of Appropriate Technology when he tried to stake a claim to the resources the county was wasting.) But we recyclers knew that there were valuable resources in the discard stream and have been proving it for over thirty years.

We were led by a simple slogan, endlessly repeated, authored by who knows who: Reduce! Reuse! Recycle! We called this slogan "the hierarchy." With the entry of the waste industry into recycling, we coined a new term for their methods: "the lowerarchy." Underlying such epithets is the hard, cold reality of value and pricing.

Pioneering recyclers lived frugally while learning how and where to connect materials with buyers. Lots of collection start-ups were in odd places like behind service stations or in schoolyards. Urban Ore started at the dump. Early on the waste managers sat back and waited for these recycling zealots to fail on their beachheads.

Some did fail, but lots of the start-ups grew, morphed, and spun off, creating a vibrant new disposal service field that steadily took away supply from the wasters. About forty years later, we've still got lots of undeveloped resources left to mine, but we've grown hugely.

Extracting Aluminum Two Ways

Urban Ore's mining metaphor fits what we were doing thirty years ago, and in a different way it still does today. For example, Dan and his coworkers extracted aluminum at the dump: ten- and twenty-yard roll-off boxes full. Not aluminum from cans, but aluminum from the industrial products that show up at landfills: windows, doors, cable, engine blocks, car parts, sheet goods. In those days, there was so much of it that the scavengers didn't always bother to pick up beverage cans.

They extracted steel, too, but the irony (no pun intended) is that in Dan's youth while saving money for college he worked for about a year in the aluminum industry extracting aluminum from bauxite. The contrast amazes us to this day.

On the one hand, picture Dan with a handful of coworkers working outside in all weather (no building!) extracting steel screws and glass and rubber and sorting the valuable bits into five or six scrap categories. "Irony" was one scrap category; that's aluminum lightly contaminated with steel. Lawn chairs are a good example: basically aluminum but with many steel pins and rivets. The different metal pieces sell at different prices.

On the other hand, picture Dan with about 1,200 coworkers feeding carbon, bauxite, and electricity into hundreds of red-hot furnace pots. These room-sized glowing tubs ran 24-7. The molten elemental aluminum sank to the bottom of the "bath." Then it was sucked into a crucible as a molten liquid and poured into molds. All this precious aluminum was extracted from a white earthen powder (bauxite) mined in the tropics and brought by big barges thousands of miles to Ohio where Dan worked.

Producing aluminum in basic reduction furnaces requires vast amounts of coal. Dan worked in the mill's carbon plant and in less than a year personally processed thousands of tons of different kinds of fossil coal into anode and cathode mixes. The recipe produced a hot black substance that looked like asphalt but smelled quite different. All that coal—as well as the coal used to produce the electricity that made the heat that drove the oxygen out of the bauxite—turned into carbon dioxide.

Extracting aluminum the Urban Ore way produces only a tiny carbon footprint.

Mind and Matter

Thirty-one years later, as we write these words, Dan has already got his long-sleeved workshirt on and has at hand his sheaf of index cards where he records our production data and his stray thoughts. In a couple of hours he'll be hard at work cashiering, restocking doors, pricing things, and working with an enthusiastic staff and thousands of drive-in customers. Urban Ore's parking lots fill up several times a day, so there is a lot of trade to manage.

FIGURE 15.1. Dan Knapp specializes in recovering and reselling secondhand doors, among many other objects and materials recovered from the deconstruction or renovation of old buildings. *Photograph by Urban Ore*

Dan is too specialized to extract much aluminum now, but our coworkers still extract tons of the metal, as well as iron, steel, copper, lead, and brass.

When working for the company, Dan does mental stuff in the mornings and early afternoons and physical stuff—real honest labor that is his main source of exercise—into the evening. Mental and physical work split around 50-50 these days. When his mental-to-physical ratio gets swamped by demands from the physical side, he stops e-mailing and puts his nose to the grindstone and his shoulder to the wheel until the underlying conditions give him time to rejoin the mental fray. The garden gets weedy, too.

On the physical side, Dan started as a generalist and has evolved into a specialist for one key commodity: doors (see figure 15.1). But both of us still know a lot about everything else we handle, from hardware to antiques. On the mental side, Dan learned business fundamentals, at first an oil-and-water situation emotionally because of the thought process he imbibed at the university that was deeply suspicious of business.

A Huge Business Today

Dan and I don't describe our chosen field as *waste management*. Waste management is a very different endeavor, related to our field only by its feedstock, although we and others in the recycling business still make diminishing use of waste management services. Our first solid waste manager called landfilling "the baseline alternative." Waste management is the competition, and it can play rough, sleazy hardball.

As businesspeople we're proud that Urban Ore is now one of thousands of recycling and reuse enterprises competing with the waste managers for the municipal supply of "discards." In fact, according to a report conducted by R. W. Beck for the National Recycling Coalition (NRC) and partially funded by the EPA, "The U.S. recycling and reuse industry consists of about 56,000 businesses that employ over 1.1 million people, generate an annual payroll of nearly $37 billion, and gross over $236 billion in annual revenues."[1] And that was in 2001; it has grown a lot bigger since then.

The supply of discards is all those gazillion tons of refined resources that years ago we mistakenly assumed were so inexhaustible that we could force people to waste them by giving them only the choice between two destructive disposal methods: burning and burying.

For Dan and me, recycling is all about saving and conserving value, getting and staying healthy, and having fun doing it.

Recycling Disposal Is Both a Service and a Process, Just Like Wasting Disposal

If you, the reader, can grasp the idea that recycling is a form of disposal, you'll be able to appreciate why—despite unfair obstacles, sharp competing practices, and regulatory barriers—recycling is slowly winning the battle over wasting. We recyclers have a big underlying structural advantage. That's because the disposal function is the heart and soul of the recycling business experience, *and we do it better.*

Let's assume that you have something in your space that you want to dispose of. If you bring it to Urban Ore and we accept it, the chances are very good that it will be conserved, upgraded, and returned to productive use. If you bring it mixed and mashed to your local transfer station, landfill, or incinerator, it will be destroyed or contaminated. Either way it will be rendered unfit for further productive use.

The service of disposal offered by recyclers and wasters alike is making your unwanted thing go away legally. We help you get it out of your life, in a sustainable fashion and without the dangers posed by landfills and incinerators.

Recycling Disposal Competes with Destructive Disposal

Destructive disposal is entirely dependent on disposal service fees both for its operating costs and its guaranteed profits. These profits are guaranteed because wasting is the only disposal service method that satisfies local governments' "sanitation" requirements.

Consider the example of a city that is pursuing the possibility of building an incinerator to take care of its waste. In order for the company to secure the investment capital to build the machine, they will need the city (or local county or even larger jurisdiction) to sign put-or-pay contracts. Once these are signed the company is guaranteed profits for twenty-five or more years. Meanwhile, the region's taxpayers will find themselves paying for this investment for that same period (or even longer if the plant later gets retrofitted to meet new regulations). These guaranteed profits offer life support for the wasting system. It's difficult to fully replace the destructive disposal industry when it is so heavily subsidized by taxpayers.

We recyclers are not asking for guaranteed profits. What we ask for is simple. We need destructive disposal to be priced at its full cost. If destructive disposal is fairly priced, it will be expensive because good-faith efforts must be made to protect the public from all the nasty byproducts (leachate from landfills, air emissions and toxic ash from incinerators). In this financial environment, genuine recyclers can work against the waste disposal price ceiling to keep our nimbler and more efficient disposal services cheaper, thus gaining us market share and thus reducing waste (see chapter 21).

Sure enough, this is what has happened as the competition has intensified. In our area during Dan's thirty-one-year reuse and recycling career, the cost for you and me to dispose of something by wasting (destructive waste disposal) has gone from $4 per cubic yard (about $15 per ton) to $126 per ton (and even higher in other transfer stations near Berkeley). That's a big disincentive to waste. All the waste haulers understand this.

Recycling and reuse companies uniquely share a common financial trait: They have two income streams. One comes from disposal service fees. The other comes from product sales. Business success depends on the balance of the two, which must pay all costs while generating reasonable profits. Business survival requires income to be greater than expenses.

Run right, almost all forms of recycling disposal can now be cheaper than wasting. For materials that have high present value, like reusable goods or deposit containers or bounty goods like cathode ray tubes, disposal service is often free or even paid for by the government or the company that accepts the discarded material. For just about everything else, disposal service is a fraction of the cost of wasting, because materials recovery generates products that are in demand and can be sold.

Besides full-cost pricing for destructive disposal, two more big challenges face clean and green recycling today. These are land availability and product quality.

The Land Question

Solid waste destruction disposal has a small urban footprint, but it requires a very large rural footprint. That's because solid waste destruction disposal mixes, mashes, and compacts materials before transporting them to rural landfills. The new country landfills have to be very big to absorb the huge volumes of compacted waste that will flow to them for many years. In North America landfills have decreased in number and increased in size.

Meanwhile, recycling and reuse require fairly large urban spaces. That's because materials have to be received, sorted in many steps, and aggregated for sale. Urban land requirements are large for recycling and larger for reuse. Composting, in our view, is a far better use of rural space than megalandfills (or, heaven forbid, incinerators).

We have long argued that adopting an airport or seaport model for resource recovery parks makes sense (see our discussion of ecoparks below), in order to take full advantage of entrepreneurial talent that is already available and hard at work. What we're after is a fusion of a secondhand shopping mall and port with an administration that is not divided against itself, as is the solid waste management profession. To date only a very few governments are following this model in their procurement processes. Instead, following the waste management model, they tend to specify more or less automated facilities, run by a single operator, that do not maximize either jobs or resource quality. This leads directly to the second problem—contaminating resources in the name of recycling.

Contaminating Resources in the Name of Recycling: A Sin Against Nature

Gradually the waste managers realized that they had to do something to meet the challenge to their business-as-usual destruction disposal coming from the grassroots and small-scale recycling and reuse operations. So they started bigger and more automated versions of the labor-intensive business models that recyclers had developed. Drop-off, curbside, and buyback operations proliferated under various brands. Cities and counties specified that they wanted a full range of recycling services, no matter who provided them. This blurred lines of demarcation that used to be clear and paved the way for the rise of big, publicly traded waste companies specializing in a new kind of "recycling."

A few years back we toured a two-acre industrial building and five-acre yard in a nearby city. The previous tenant was a huge publicly traded waste company. The building and land was used to beneficiate glass (sort, clean, and crush it) sourced in-house from their operations at all their franchises in California. The glass material was extremely dirty due to automated upstream handling practices. Nutrient-rich food was the primary contaminant. Rats, cats, bugs, and seagulls proliferated in the piles and spread to adjoining businesses. After exhausting other remedies, the nearby landowners banded together and got the publicly traded company evicted. Now a new owner has cleaned up and put the property back on the market, and a consortium of clean recycling partners is angling to buy or lease it.

Alternative Daily Cover

The problem of contaminated resources has reached a kind of nadir in California, where since 1994 it has been legal to get "diversion" credit for manufacturing a finely shredded form of rubbish called alternative daily cover (ADC). In this way, the waste industry subverted landmark legislation that in 1989 mandated what people thought was a 50 percent recycling goal. It was no such thing after the waste managers reworked it to create the diversion loophole. As I write this, ADC has surpassed clean composting as the biggest contributor to diversion in California cities, even though its only use is to cover compacted garbage, which is then buried the next day. ADC is clearly wasting, especially when you consider that much of it could have been reused if there was any competition allowed or encouraged on the same property as the processor shredding material for ADC. For example, the Zanker Road Landfill in San Jose is recovering over 90 percent of the C&D debris it receives.[2]

There are more examples, but the point is this: We believe that those of us for clean recycling must stand firmly in its defense and must demand that the dirty recyclers and purveyors of the lowerarchy give way to more sustainable practices. We must develop ways of policing ourselves against dumbed-down quality or we risk being caught up in a backlash and a verification of cultural prejudices against our kind that are already part of the societal fabric. We must fight the primitive and unnecessary act of making some people into a caste of "untouchables" because they handle discarded materials. These old and stubborn cultural prejudices reinforce wasting, but we can change them with good, sensible, safe, legal, and profitable recycling practices. As Helen Spiegelman puts it in her essay in chapter 18, we have to "occupy recycling."

Recycling Is a Powerful Tonic for What Ails the American Economy

True blue recyclers specialize in products without pollution. Not relatively worthless products like buried garbage or ashes or food-contaminated glass, but valuable products like high-quality topsoil, usable doors and windows, garden lumber, and clean feedstocks for industry like graded metals and ceramic-free, color-sorted glass cullet. These are the products we sell for good prices to local markets. Selling locally at a fair price completes the disposal loop where it matters most.

It takes lots of people to sort and handle all these discards. Some operations use more labor per unit of output than others. Urban Ore, for example, is very labor-intensive, comparable among business types to a law office or a baseball team. Urban Ore is knowledge-intensive, like a university. Recycling uses intellect and physical strength in equal measure. A ceramics recycler is more capital-intensive than a reuser, but they are both relatively labor-intensive, which means they create paying work for people to do: kids, dads, moms, all kinds of people who enjoy materials and enjoy people and enjoy working hard and trading stuff. These folks are legion.

So let's make our products and let's sell them to people who need and use them. Let's empower and supply a rebuilding boom. Let's provide an outlet for deconstruction (not demolition) companies and provide materials and products for green architecture. Let's conserve and enhance value. Let's become indispensable and sanitary. Let's build resource recovery parks and run them with resource development authorities. Let's stop all landfill expansions. Let's not subsidize wasting anymore. Let's get all organics out of landfill and stop burning what we can better compost and return to the soil.

Let's go all the way to zero waste, and the sooner the better. To get there we need to separate all discards into twelve master categories and then handle these in ecoparks.

The Twelve Master Categories of Discards

There are twelve master commodity sets in the recycling industry today, and materials recovery businesses compete for supply every day in each and every one. Many recovery businesses handle more than one commodity, often in related clusters. In general, the more a master category's resources can be divided into subcategories, the more upgrading they can receive. Then they generate more in sales, and this cash flows throughout the economic system.

1. **Reusable goods**, including intact or repairable manufactured objects such as home or industrial appliances, household goods, and clothing; intact materials in demolition debris, such as lumber or bricks; building materials such as doors, windows, cabinets, and sinks; business supplies and equipment; lighting fixtures; and any other items that can be repaired, repurposed, or used again as is.
2. **Metals**, including both ferrous and nonferrous objects such as cans, parts from abandoned vehicles, plumbing, fences, metal doors and screens, wire, and any other discarded metal objects.
3. **Glass**, including glass beverage containers, drinking glasses, shower doors, window glass, and other objects made of glass.
4. **Paper**, including newsprint, office paper, ledger paper, computer paper, corrugated cardboard, and mixed paper.
5. **Polymers**, including beverage containers, plastic bags, plastic packaging, tires, and plastic cases of consumer goods such as telephones or electronic equipment.
6. **Textiles**, including nonreusable clothing, upholstery, and pieces of fabric.

7. **Chemicals,** including recyclable or reusable solvents, paints, motor oil, and lubricants.
8. **Wood,** including nonreusable lumber, pallets, furniture, and cabinets.
9. **Plant debris,** including leaves, cuttings, and trimmings from trees, shrubs, and grass.
10. **Putrescibles,** including garbage, offal, and debris from animals, fruits, and vegetables.
11. **Soils,** including excavation soils from barren or developed land and excess soils from yards.
12. **Ceramics,** including rock, porcelain, concrete, and nonreusable brick.

Frequently Asked Questions About Urban Ore Ecoparks

At Urban Ore we have designed zero waste ecoparks to handle these twelve master categories and maximize their reintroduction into the (preferably local) economy (see figure 15.2). Here are the questions we most frequently answer about them.

What is the science of zero waste?

Zero waste is recovering the twelve master material categories listed above fully from the discard supply.

What about the residuals from this system?

Jurisdictions can attack unrecyclable residues with EPR and landfill bans. Another way to tackle this problem is Step 8 in the ten steps to zero waste outlined in chapter 2, building a residual separation and research facility in front of the interim landfill.

What is a zero waste ecopark?

A zero waste ecopark is a specialized port where discarded resources are collected, processed, and fed back into commerce.

Why build zero waste ecoparks?

Zero waste ecoparks are powerful economic engines, harvesting cash from disposal service fees and product sales. Reuse is especially labor-intensive, which means it is employment-intensive. Recycling is next, and then composting. But even high-tech composting gets labor-intensive when the products are used by the landscaping and agricultural industries.

There is a big opportunity right now for entrepreneurial ventures to expand materials recovery, because so many good talented people are available to work. Besides, the alternative is mixed-waste landfilling or incineration—sunset industries. Both industries pollute and waste resources. Neither is sustainable.

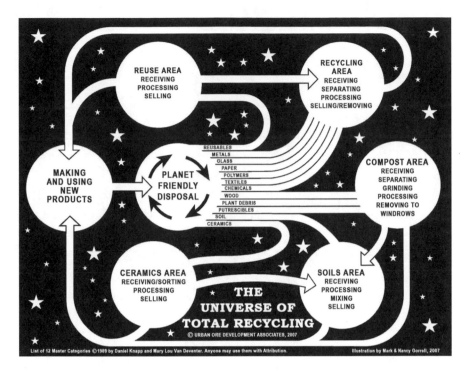

FIGURE 15.2. The twelve master categories of discards at an ecopark. *Photograph by Urban Ore*

Can existing discard management facilities be modified to become zero waste ecoparks?

Yes. Niche reusers, recyclers, and composters already cluster naturally due to market forces. Since they compete with wasting industries for supply, they like to be close to waste transfer stations.

Can zero waste ecoparks be purpose-built?

Yes. Such ecoparks will be magnets for high-quality resource streams. Zero waste ecoparks can be designed for business clusters that handle reusables first, recyclables next, and then clean compostables—with wasting as a last, most expensive resort.

Disposal service fees for materials conservation venues should be lower than wasting venues. Haulers will sort for cheap disposal. They've been paid for wasting, but the difference between recycling and not-recycling is cash left in the pocket.

What is the best scale for zero waste ecoparks?

Any scale is a good scale. We live in a near–zero waste household. Our business is a near–zero waste business. If our local transfer station complex were rebuilt to operate more like a port than a parking lot, our company and many others that reside in it would all benefit with greatly increased throughput. Our faithful customers and vendors would like it, too.

How is zero waste incentivized within the ecopark?

Open facades, airport-style unloading, assisted unloading, price breaks for quality materials, payment for some, deposit redemption, and legal disposal. The order of access should be reuse, then recycling, then composting, then wasting. Disposal service prices increase farther in. Cheapest or most remunerative clustered toward the front. Minimize on-site sorting with rates.

Who will own these zero waste ecoparks, and for whose benefit?

Smart municipalities will create airportlike special purpose districts to develop and manage these facilities. Their goal should be to provide maximum public benefit by developing an ecology of recycling commerce.

Who will manage these zero waste ecoparks, and to what ends?

You, the reader, are the best one to answer this question. Just do it.

Waste Is a Social Issue First,
a Market Issue Second

By Eric Lombardi

Another visionary who believes in the zero waste ecopark approach is Eric Lombardi. He is currently the executive director of Eco-Cycle (www.ecocycle .org), and since 1980 he has special- ized in resource conservation, social enterprise development, and public- private project development. Eco-Cycle, founded in 1976, is a national pioneer in the recycling industry and has grown under Lombardi's tenure (since 1989) to become the largest community-based recycling organization in the United

ERIC LOMBARDI

States, with a staff of seventy and processing of over fifty-five thousand tons of diverse recycled materials per year in 2011. Lombardi is recognized as an authority on developing comprehensive community-based programs and is often a keynote speaker and consultant on the social and technical aspects of creating a "zero waste—or darn near" society.

Lombardi has designed and built numerous community recycling centers in Boulder County, Colorado; built and operated the first Center for Hard-to-Recycle Materials (CHaRM) in the United States; was a USAID consultant for the creation of a paper recycling program in eastern Romania; and led the design effort of the new material recovery facility (MRF) in Boulder County that was declared "the nicest MRF I've ever seen in the world" by George Weyerhaeuser, the vice president of paper recycling for the corporation of the same name (see figure 16.2).

Lombardi was invited to the Clinton White House in 1998 as one of the top one hundred US recyclers to advise on national recycling policy. He cofounded and served as the board president of the national GrassRoots Recycling Network (GRRN) and is a cofounder of the global Zero Waste International Alliance (ZWIA). He served from 1997 to 2004 on the National Recycling Coalition's (NRC's) Policy

FIGURE 16.1. The Boulder County public MRF designed and operated by Eco-Cycle in Boulder, Colorado. *Photograph by Eco-Cycle*

Work Group and is a past board member of the Colorado Association for Recycling (CAFR). Lombardi was featured in a 2008 Newsweek *article "10 Fixes for the Planet" in which one of the fixes was zero waste.*

The history of how the world deals with its trash has evolved rapidly since the early 1990s—from open dumps on the edge of town to zero waste community programs that recover nearly all our discards for recycling, composting, or reuse. This amazing success story has happened due to our increasing knowledge and concern about the impacts of waste "disposal" on our collective public health and environmental quality. This beneficial social change did *not* happen because the free market sent financial signals through its invisible hand into our homes and businesses that made it more "profitable" for us all to recycle—in fact, it's tough to dispose of waste more cheaply than digging a hole in the ground and burying it. So why then is there a global revolution in how we are all taking out the trash?

The answer to that question is, as they say, boring but important. It is because the macroeconomic concepts of environmental externalities and market failures have finally caught up with the waste management industry. Basically, this means that the pollution from landfills and incinerators that is being released into the environment in the form of toxic air emissions and leachate are not being paid for by the polluter, thus this bad behavior has

FIGURE 16.2. Two of the separation lines at the Boulder County/Eco-Cycle facility in Boulder, Colorado. *Photograph by Eco-Cycle*

continued even though there are real economic impacts being paid by nonpolluting bystanders.

A personal experience serves to show how this corrupted free market system has impacted my town: The City of Boulder and its residents used a privately owned landfill near town until it was closed by government action because it was polluting the groundwater of a nearby community. The landfill was immediately designated a federal Superfund site, and over $13 million was spent over many years cleaning up the groundwater under the landfill. Half of that money was paid by the landfill owner and half by the city (the general public). Not long ago the EPA ruled that the groundwater cleanup phase can stop but that the city would be responsible for keeping rainwater from entering that old 160-acre landfill "in perpetuity"—in other words, forever. So the question now is, "How much does forever cost per ton?" Because unless we can answer that, we will never know the true cost of landfilling at that site.

Yet local recycling programs were criticized during this period of artificially cheap landfilling as being "too expensive," and that perception continues into today. We need to ask, "Too expensive compared to what?" The true landfill costs in Boulder have never been known and will now include ongoing legacy costs of an unknown but significant amount that will be paid for by future bystanders who had nothing to do with creating this toxic site. I call this "dishonest economics," or the polluter-doesn't-pay system, and it is time to

FIGURE 16.3. Clean organic materials are collected for composting.
Photograph by Eco–Cycle

dispel the myth that the market prices of landfilling or incineration of waste accurately reflect their environmental impacts. Yet the opposite is true for zero waste systems such as recycling, composting, and reuse (see figures 16.2 and 16.3)—all environmental costs are transparent and there are no legacy costs.

So How Does One Change a Dishonest Economic System?

This is a huge question, and the best minds in the world have been attempting to create a new "ecological economics" for many decades with little success. As a social scientist, I believe there is another way forward to handle the problem. Let me submit to you my belief that slavery in America did not end because President Lincoln was shown a spreadsheet that told him it was bad for the national economy. Rather, the glaring evils of human bondage were sufficiently obvious at that point in history to simply elevate the debate above economics and create a new social paradigm on how life, business, economics, and everything would forever be changed in a world without slaves. I believe we are at a similar point in history where the evils of environmental destruction are now so obvious that we must lift the debate above the economic discussion and create the new paradigm for how we live more gently on the earth so that we stop ravaging nature and our limited resource base so that future generations have a chance to live in clean, harmonious societies. In the nineteenth century, people were certain that the end of slavery meant the end of prosperity, but before long we figured it out and the world became a better place. Today many people can't imagine prosperity without environmental degradation, but many more people are already busy creating a new green economic world.

If we are to rise above dishonest economics and start a new dialogue beyond economics, does that mean socialism and the end of the free enterprise system? Absolutely not, but it does mean that some business will be done differently. What stays the same is the high risk–high reward marketplace where companies like Apple compete and bring high-tech gifts to the world for which we are grateful. I am not talking about that world, but rather am referring to the low risk–low reward marketplace, such as community electric utilities or community waste management systems. It is this marketplace that we need to transform so that a true and legal double-bottom-line (DBL) approach can thrive and create for us social and environmental benefits while generating a financial profit of limited scope. This approach has a name—it is called social enterprise (SE), and it is the next big thing in social change movements.

SE in the zero waste field is a perfect fit for many reasons. First and foremost is the "mission" focus as the first priority of every SE. In fact, that is probably the defining difference between a purely for-profit enterprise and an SE. A second reason is that the transition of the waste management industry into a "resource management" industry will require a business approach with all the market powers of efficiency, accountability, and profit motive at play. Just because a zero waste social enterprise (ZWSE) may find its profits limited to something around the "public utility rate" of 10 percent, it is still true that 10 percent of a large number is in itself a large number, and the future will continue to offer the opportunity for many people to create new SEs in the waste industry and become zero waste millionaires (but perhaps not billionaires!).

Zero Waste Capitalism

Now you can see why I was introduced during the Italian zero waste Tuscany tour 2009 as the "zero waste capitalist from America." I advocate for a brand of market activity that will build and operate the green community infrastructure, like zero waste city systems, that we so desperately need—and I believe SE could become the emerging "true, legal, and accountable" DBL business vehicle the world needs. Since the early 2000s, the United Kingdom has experimented with a legal entity called the community interest company (CIC); in the United States we now have the low-profit limited liability company (L3C). Both of these legal entities are revolutionary in their potential. In Italy there is also a long history of community cooperatives, and there is a University of Bologna research center SE known as the Associazione Italiana per la promozione della Cultura della Cooperazione e del Non-Profit (AICCON).

The Government's Role

The government must play an important role here and do two things to support the success of future ZWSEs: 1) make it a requirement that every home and business separate their discards into at least three categories—recyclables, compostables, and residuals; and 2) consider taking public ownership of all discards within their jurisdiction and "flow" these materials to their SE public-private partners through either competitive or negotiated contracts. Neither of these government actions requires money, only vision and courage. Once the government fulfills their role of committing and planning for the resource management transition in their community, then the private-sector SEs can come forward with the capital and experience needed for success.

The existing, status quo waste management industry is a single-bottom-line (SBL) activity that is dependent upon old technologies that destroy, rather than conserve, the natural resources in our communities. That old path will conflict with our new vision of ZWSEs running DBL companies because our new system comes to town and demands that they change or die. But there is also an opportunity here; because the transition is going to take time, every SBL waste company could choose to transform itself into a DBL SE. The fact is that they are experts in collecting and hauling discards, which is an important piece of the zero waste city system, so there is no reason why a forward-looking waste company couldn't evolve by changing its business model by becoming a partner and service provider in the future green community systems. Just look to the ex-garbage, now zero waste company named Recology in San Francisco for a large-scale example of this idea.

The Battle over Discards

But let's not kid ourselves. As I write this in the year 2012 there is a war going on in the alleyways of the world over access to the discarded resources of the community. The waste companies (wasters) want our discards to stay all mixed up so that they can cheaply collect and then bury or burn the material in their landfills and incinerators. On the other hand, the recyclers and composters want the discards nicely sorted and brought to their recycling and composting facilities. These are two business models competing over the same feedstock for their success. But there is no comparing which system creates more benefit for the community, and even the wasters will admit that a "certain amount of recycling is good." But how much do they think is good? Any amount that won't seriously threaten their immense profits is OK, and I'm guessing that they can live with a recycling rate up to 50 percent—but beyond that they will struggle to afford to build and operate large disposal facilities like landfills and incinerators, which is the source of their largest profits; the world will see a real struggle as the old dinosaurs slowly die off.

A Total Commitment

When the zero waste community proclaims that we are not simply a puzzle piece in an integrated solid waste management (ISWM) plan, what we mean is that the zero waste path is a *total* commitment toward achieving zero, even if we never actually get there. "Now we know"—a favorite phrase of mine in this amazing age of the Internet—that even a major city can achieve an 80 percent recovery rate, and small towns have reached 90 percent. Zero waste is a journey, not a destination, and every decision we make in industry, government, and our homes gets us one step closer or further from our goal of *zero* waste. It is a goal worthy of a changing world.

One-Stop Dropping

In Boulder, Eco-Cycle has endeavored to put as many recycling and reuse activities on one site as we can. I would like to see this idea further developed into what I call an "Eco-Cycle zero waste park." As explained in chapter 2 this would offer a "one-stop dropping" consortium in which the "big six" zero waste facilities (a composting facility, an MRF, a hard-to-recycle facility, reuse and repair facility, a C&D facility, and a residual separation and research facility) that every community needs are offered at one site. Obviously, this would need zero waste consultants and local entrepreneurs to cobble together such a package and offer it to a local community instead of the one-stop shopping offered by mass burn incinerators. See figure 2.8 for a schematic view of such a facility.

The New Peace Movement

A final word about why we should focus on shaking up the waste industry; another favorite phrase of mine fits here: "Why bother?" Many intelligent forecasters, including Amory Lovins of the Rocky Mountain Institute and the US Department of Defense, have said that the primary source of conflict on the planet in the twenty-first century is going to be in gaining access to a dwindling amount of natural resources. This is clearly happening in Africa over access to minerals, in Indonesia over access to timber, in many regions over access to oil, and now conflicts over access to water are beginning to emerge. These resource wars have, in my opinion, begun, and it is clearly now an ethical and moral issue to waste something casually after blood has been spilled over it. The situation is going to get worse before it gets better—and so, in a broad worldview, zero waste has become the new *peace movement* for us all to support.

Gibsons Resource Recovery Center: Zero Waste from the Bottom Up

By Buddy Boyd

Buddy Boyd has had a fascinating and varied working career. He started off as a truck driver for Canadian Pacific and quickly rose in the company to become manager for the Vancouver, Calgary, Edmonton, and Winnipeg area. After becoming the Toronto area manager, he moved west and worked for a waste disposal company (Smithrite) in Vancouver; then he owned a record store (Breeze Records) and then a portable sawmill business (known as the Logbutcher). He built his own house from the trees on his land on Gambier Island. In 2003, he set up Gibsons Resource Recovery Center on the Sunshine Coast of British Columbia, which he says, "is without a doubt the best place I have ever lived, and I am trying to make it even better."

Boyd is a former member of the board of directors for the Recycling Council of British Columbia (RCBC[1]) and current member of their policy committee. He is also on the planning committee for the Zero Waste International Alliance (ZWIA).[2]

BUDDY BOYD AND BARB HETHERINGTON

Gibsons Resource Recovery Center (formerly called Gibsons Recycling Depot) is the largest non–government-funded operation of its kind in British Columbia. It has never received any funding from taxpayers and instead generates its own revenue from the fees it charges for certain materials and from the sale of clean recyclables and other objects that we (like Daniel Knapp at Urban Ore) "mine" from the discard stream that comes through our gates. My life partner, Barb Hetherington, and I see garbage differently from most people. We reclaim resources from "garbage" and put them back into the community, not into the ground. Our facility currently employs thirteen people. (A videotape featuring the Gibsons Resource Recovery Center can be viewed online,[3] as can a trailer for the related film *The Clean Bin Project*.)[4]

Our center accepts all kinds of household discards and used items to be resold or remade into new products. In all we accept twenty-one different items ranging from textiles to electronic equipment. Some items are received without a fee; others have a fee charged. The practicalities are described in more detail in chapter 9.

In 2010, I was invited to speak at an international conference on zero waste in Florianópolis, Brazil. There I heard all sorts of information about what a zero waste community looks like, and it turns out that we've moved a long way in that direction at our center. But there are many more things I would like to do.

I would like our operation to become a community meeting place to encourage people to come out and spend some time together. Operations of this kind in Sweden boast attractions like a kids' play area and a restaurant. I would like to incorporate reuse and repair as well as composting and an education center at our facility. Education is the crucial step to creating a zero waste community locally.

Sadly, there are no government programs in British Columbia to provide funding to build operations like ours or help purchase the equipment that could keep more material from going into landfills. Instead of government money going into genuine recycling operations like ours, it is being frittered away building hugely expensive resource-destroying incinerators in the metropolitan Vancouver area.

Some new words have entered into the solid waste conversations here in British Columbia like *sustainability, resource recovery, extended producer responsibility (EPR)*, and *zero waste*. Hallelujah, we thought: Finally some positive ideas on how waste can be viewed as a resource—not something to get rid of. We anticipated that zero waste policies would show consumers exactly how to transition from wasting to zero waste. Yes indeed, exciting times. Many of us who support the genuine zero waste movement joined committees, recycling boards, and volunteered for countless hours to become involved in helping our province to become a more sustainable place. We felt very positive about the future and how the principles of zero waste would become the cornerstone for a new era where we would move from wasting to sensible discard management both from the top down and the bottom up.

But we didn't realize how easily both the common sense and the wisdom contained in the zero waste message could be undermined by those who wish to make giant profits out of continuing to "dispose of waste." Before we knew it, the recycling, reuse, repair, and zero waste grassroots organizations, groups, and committees we belong to were being invaded by "sustainability" experts and "consultants" who always were part of the wasting lobby but simply had changed their titles and labels. We have seen rebranding and greenwashing on steroids.

Our very own British Columbia Ministry of the Environment has formally recognized burning garbage as a form of "renewable energy." Imagine that: Resource destruction is now being billed as resource recovery! And the huge "waste of energy" that incineration represents (see Jeffrey Morris in chapter 21) is billed as "waste-to-energy"!

Instead of the efforts like ours to carefully separate materials into many different categories, thereby enhancing their value, we now see our government encouraging the commingling of all recyclables together in single-bin collection systems. This despite the fact that it is abundantly clear that citizens are more than willing to separate into more categories if they know that by doing so their children and grandchildren will have a better future. As Helen Spiegelman and others have pointed out, the commingling of all these materials is likely to produce contaminated materials of such low value that they either have to be shipped overseas or sent to an incinerator. Even our EPR programs have incineration as an acceptable fate for some of the materials captured. But never mind, they say; this is all part of their version of "resource recovery."

The epitome of this British Columbia doublespeak is that Metro Vancouver has created the Zero Waste Challenge[5, 6] but this will include building a new $500 million waste-to-energy (WTE) facility to burn five hundred thousand tons of waste a year. And that's not the end of it. They are also looking for an experimental technology to burn another one hundred thousand tons per year. All of this is proposed by Metro's Zero Waste Challenge experts. In other words, the zero waste message has been turned on its head here in British Columbia. I am not making this up. You can read the gory details in a *Globe and Mail* article published on January 3, 2012. The title gives you the flavor—"Incineration Firms Vying to Burn Vancouver's Trash":

> The competition is fierce because of Vancouver's reputation as an environmentally progressive, livable city. Local politicians say that aspect is something that will enhance the reputation of whichever system and company is chosen for the contract to dispose of the region's 500,000 tonnes of garbage. That will be like an advertisement to thousands of cities around North America who are being forced to contemplate a future where landfills are no longer an option.
>
> "They want to do business with us in Vancouver because we're a leader in green," said Metro Vancouver chair Greg Moore.[7]

One thing is for sure: Zero waste is nowhere near being achieved in either Vancouver or at the provincial level in British Columbia. The whole process seems rigged in favor of the continuation of wasting and supporting the status quo. And with extensive greenwashing and rebranding at every level of government, zero waste and sustainability are little more then an opportunity to create new staff positions in government. It is the equivalent of "talking the walk."

As the owners and operators of Gibsons Resource Recovery Center in our community, Barb and I are not welcomed by the waste companies and their friends in government. They see what we are doing as a threat. Their message is that zero waste is something worthwhile but obtainable only in the distant future. Meanwhile "sustainability managers" at the government level (formerly solid waste coordinators) continue to hire solid waste consultants and experts to keep developing the community solid waste plans—which in turn support more studying, reviewing, and consulting to keep zero waste under review, while the real game in town is making huge profits for a few individuals by wasting our resources in landfills and incinerators.

The Need for Real Political Leadership

Is zero waste achievable? Absolutely! But what we need most in British Columbia is real political leadership from those whom we have entrusted to lead us, like the leaders in San Francisco; many towns in Italy; Spain; and Kamikatsu, Japan.[8] One of my favorite bumper stickers is the one that says,

"When the people lead eventually the leaders will follow." One experience in my life showed me what real leadership is about.

In 1985, when I had my record store in a Vancouver suburb, Vancouver was hosting the 1986 World Exposition on Transportation and Communication. The Expo theme was "World in Motion." My neighbor, Rick Hansen, was paralyzed with a spinal cord injury and decided to wheelchair around the world, calling his journey "Man in Motion," to raise money for spinal cord research.

I was operating my record store during the day and driving garbage trucks at night. I approached Hansen and asked him what he was going to do for "tunes" for his daunting journey. After a brief discussion with his team, we became the official supplier of music for the "Rick Hansen Man in Motion World Tour."

Most Canadians know that Hansen successfully completed his world tour. He has shown that anyone can do anything if they put their mind to it. That is what I call real leadership—leadership by doing, not just talking about it. Needless to say, Hansen has been a huge inspiration for me to overcome much simpler obstacles. These have included milling my own trees into lumber to build my own house on a remote island. Hansen's example has also served me well in overcoming all the hurdles and obstacles when it comes to managing our discard stream as a resource instead of waste.

Lessons from My Life

I have tried to analyze the kind of life experiences I have had that have led me into this kind of work. I am not religious, but I am spiritual. Living in the bush on Gambier Island for over fifteen years—with no services and with only 120 neighbors—shapes a guy pretty well. Heat, water, food, and waste all take on an entirely different priority. I strongly recommend everyone drop out at least once in his or her life. You get time to think, to breathe. Time to find one's real values. Time to distinguish between what we need and what others (particularly the advertisers in our consumer society) persuade us that we need. We can't make a difference if we let money or ego drive us. We are a family of earthlings who have lost the ability for the most part to think for ourselves on the one hand and work cooperatively together on the other.

Gibsons Resource Recovery Center is the culmination of all those choices, lucky breaks, and happenstance coming together at the same time. Barb and I are just the conduits. Being curious by nature, we just connected the dots. Anything is possible if you just try. The survival of humanity will only be possible through a team effort from the bottom up, not top down.

My hope is that grassroots activists in Vancouver can delay the building of new incinerators long enough for more resource recovery centers like ours to spread like mushrooms around British Columbia.

Multimaterial Curbside Recycling and Producer Responsibility

By Helen Spiegelman

Helen Spiegelman combines a passionate commitment to local grassroots organiz-ing with a keen global overview of the waste issue. She is the cofounder of Zero Waste Vancouver. That initiative led to a campaign from 2007 to 2010 against Metro Vancouver's proposal to include incinerators in their waste management plan (this threat is not dead but at bay). Now she is active in what is being called the Campaign for Real Recycling, a local   **HELEN SPIEGELMAN** *initiative inspired by the initiative with the same name in the United Kingdom.[1] This campaign is calling for deposits on milk containers and an end to single-stream recycling. In 2003, Spiegelman cofounded the Product Policy Institute (PPI) with Bill Sheehan (whom we hear from in chapter 19).*

The zero waste movement is a reaction to the long period of socially sanctioned wasting that characterized the twentieth century. But while it's easy to reject the idea of waste, it's harder to root out its pervasive political and institutional basis—and harder still to hold on to the zero waste agenda when powerful forces are hijacking it for their own purposes.

This chapter picks up on an argument raised by Samantha MacBride in her 2012 book *Recycling Reconsidered* that corporate interests have had an inordinate influence in shaping recycling policy and programs for the past forty years.[2] It frankly acknowledges the contradictions in the now-prevailing recycling system: multimaterial curbside collection programs operated by the waste industry. It argues that the new policy of extended producer responsibil-ity (EPR) presents us with a crisis in the development of a sound waste policy: a threat of complete corporate control combined with a brief opportunity to take back control and reshape policy to serve the public interest. It closes with breaking developments in British Columbia that could open a window on what new opportunities might look like.

A Bit of History

The twentieth century marks a period when an elaborate civic and industrial apparatus evolved to support wasting. This apparatus is referred to by its practitioners as the municipal waste management system. It is operated jointly by municipalities and the waste management industry in a close working relationship, where local governments are often found to be acting simultaneously as contractors, competitors, and regulators. This relationship has been referred to as the municipal industrial complex.[3] The most visible components of this apparatus—ubiquitous street litter bins and convenient weekly collection and disposal of refuse—are now viewed by the public as a necessity, like sewers, fire departments, and hospitals.

Waste management was initially undertaken by local governments to safeguard public health and safety.[4] But paradoxically, municipal landfills and incinerators—even the packer trucks dripping hopper juice on public streets and alleys—have themselves become threats to public health and safety. And the threats are not just local but global. The squandering of resources facilitated by public waste management policy and practices is one of the major drivers of climate change at both the front-end and the back-end of the product life cycle.[5]

In the 1980s, the movement that anticipated today's zero waste movement mounted a challenge to the municipal waste management system by rejecting the use of incinerators and landfills to handle society's burgeoning quantities of municipal waste.[6] The reformers were astonishingly successful. As Eric Lombardi has pointed out elsewhere in this book, recycling was integrated into the municipal waste management system, alongside landfilling and incineration. But because recycling programs were part of the waste management system, they were increasingly shaped by the realities of waste handling, and not necessarily by broader environmental, social, and economic goals.

The first casualty was recycling opportunities for small local businesses. The introduction of municipal recycling programs put many small independent recyclers out of business. In my neighborhood, a local recycler overnight lost all his accounts picking up newspaper from homeowners for free and selling it to local paper mills and brokers. Now the only sustainable business opportunity is through contracts with local governments, competing against the big garbage companies.

The second casualty was the standard of service. In communities across North America, source-separated recycling collection has been abandoned in favor of single-stream collection. This brings cost savings for haulers but reduces the value of the collected materials.[7]

As a result, recycling as it is practiced within the prevailing waste management system does the same thing that incineration does: destroys resources and opportunities. Materials that could have built the domestic recycling industry are downcycled. The mixed materials are hauled in bulk, crudely processed, and exported to distant markets instead of nourishing new local enterprises. More value is squandered than is conserved.

As this current version of recycling has grown, big waste haulers have come to control a larger and larger portion of the total supply of recyclable materials, and the materials that they are bringing to market are less valuable. Domestic paper mills and plastic processors have the choice of absorbing the cost of working with lower quality feedstock or seeing it go to offshore competitors.

Along with loss of quality, these programs are not delivering the quantity of material that manufacturers have said they need, particularly paper and plastic. Even in the most highly developed municipal recycling programs, upwards of 50 percent of many potentially recyclable materials are going to the disposal stream rather than the recycling stream.[8]

A growing body of evidence[9] shows that single-stream, multimaterial curbside recycling is failing to achieve the promise envisioned by recycling advocates in the 1980s and by the contributors to this book today. If recycling is to be, as zero wasters envision, a key component of a new zero waste society for the twenty-first century, we need to see that it is done right.

Again, looking closely at history is a good place to start. How many know the story of the involvement of the beverage industry in supporting and shaping multimaterial curbside recycling?

In *Recycling Reconsidered*, MacBride raised the question: Why have our curbside recycling programs targeted materials that have little inherent commodity value and underdeveloped recycling industries (glass and plastic) but ignored discards (textiles, for instance, and compostable organics) that could more easily be diverted from disposal and bring quicker and greater public benefit in waste reduction? Who defined the basket of goods targeted by multimaterial municipal recycling programs? (The basket also includes paper, for reasons that will be discussed later.)

In answer, MacBride documents a project launched in 1970, right after Earth Day, by a group called the Business Environmental Action Coalition

Committee (BEACC). BEACC's members included Coca-Cola, the Aluminum Association, and 7 Up, among others. Through BEACC, corporations provided financial support to a grassroots environmental organization that lobbied New York City to provide a curbside recycling program for glass and aluminum beverage containers. MacBride posited in her book, with substantiating quotations from participating corporate executives, that this was a conscious industry strategy to deflect a looming threat of producer-focused waste reduction mandates—including mandatory deposit/return programs, recycled content requirements, and bans.

Fifteen years later, the identical strategy was played out in Ontario.[10, 11] Coca-Cola and PepsiCo and Alcan formed an association (Ontario Multi-Material Recycling Inc. [OMMRI]) and brought some money to the table. In coordination with key grassroots environmental organizations, the Ontario beverage industry arranged to inoculate itself against a full-blown case of producer responsibility by paying a small portion of the costs of setting up a universal blue box curbside recycling program for bottles and cans and paper. In exchange, beverage producers were relieved of a legislated requirement to sell a quota of their products in refillable bottles.

In 2010 Coke ran the same play in Vermont. The company's lobbyists found four sponsors for a bill[12] that would have rescinded the state's (producer-focused) beverage container deposit-return system in exchange for a statewide (municipal-focused) multimaterial recycling program. The bill specified that producers would cover 100 percent of the program's operating costs. They even appropriated the term "EPR" in the title of the bill. The initiative failed because giving up the deposit system was a deal breaker.

The latest location in this ongoing industry campaign is Minnesota. An organization named Recycling Reinvented[13] has been formed to support "EPR." It is working to build "a coalition of supporters from the public, private and nonprofit sectors." Under the EPR banner, the coalition will push for a statewide, producer-financed, multimaterial municipal curbside recycling program. The Recycling Reinvented board is headlined by Robert F. Kennedy, Jr. (Keeper Springs Water) and Kim Jeffery (Nestlé Waters). These two corporate executives teamed up in 2009 to fight the expansion of New York's bottle bill to include bottled water.[14]

Do We See a Pattern Here?

In the wry words of Canadian Guy Crittenden, writing about the Ontario experience: "The Blue Box on your front porch wasn't dreamed up by government officials. Or inspired by grassroots environmentalists. The soft drink industry and its packaging suppliers brought in the Blue Box to serve a common corporate agenda—thwarting government legislation that would have foiled their plans to bury the refillable bottle in the junk heap of history."[15] So the question has to be asked: Can a recycling program designed by producers of throwaway products and operated by the waste management system be the foundation of the recycling economy hoped for by the zero waste movement? I think not.

Two recent reports provide insights on the prospects for job growth through expanded recycling. The first study, carried out by Tellus Institute and Sound Resource Management Group, describes the employment benefits of a "green case" scenario where 75 percent of municipal discards are recovered through recycling and composting.[16] This scenario results in a total of 2.3 million jobs— almost twice as many jobs as we have now with half the "diversion" rate. The reason is that recycling is more labor-intensive than landfilling or incineration.

But the report's projection rests on a critical assumption, mentioned almost parenthetically: "that the materials recovered through recycling remain in the U.S. and are utilized as inputs by domestic manufacturers." The report acknowledges that this is not the case now but provides no indication of how this situation will be changed.

The report also acknowledges, again parenthetically, that jobs in recycling collection will actually decline "as single-stream recycling continues to grow." Again, there are no recommendations that this trend be reversed by taking a stand against single-stream recycling, even though this would also help prevent jobs from going overseas.

The second recent report was issued by the Container Recycling Institute (CRI) in 2012.[17] The CRI study was coauthored by Jeffrey Morris (who also contributed to the Tellus/Sound Resource Management Group report) and Canadian consultant Clarissa Morawski. It looked more closely at recycling. The goal of the study was to measure the impacts on domestic jobs from increased recycling of beverage containers through container deposit-return (CDR) systems compared to curbside recycling or landfill disposal.

The key findings were that CDR systems create between eleven and thirty-eight times more jobs than a curbside recycling program for beverage containers. The principal reason is that the CDR system has triple the capture rate (percent of total available material that is actually recovered) compared to curbside recycling. Containers are handled more carefully under a CDR system, making the industry more labor-intensive, providing 1.5 to 5 times more jobs than curbside programs in collection, sorting, and transporting. These are local jobs that can't be outsourced overseas. Container deposits are the EPR policy most feared by the beverage industry.

These two studies come at a time when the zero waste movement is having an internal debate over EPR policy. Early advocates of EPR emphasized the need to address the waste problem at the beginning of the pipe. They saw EPR as a waste prevention approach. Critics of EPR within the zero waste movement emphasize the risk that EPR will hand over control of recycling to corporations, eliminating local control and jobs. How can these views be reconciled?

EPR introduces for the first time the notion that a producer has "responsibility." This is a powerful instrument for public control, but only if it is the public, not producers, who define what the objectives are, what the responsibilities are, and how we know when they have achieved them.

From the perspective of an early advocate of EPR, I offer an initial list of requirements that must be included within an EPR regime:

1. Conduct an economic impact analysis that demonstrates how the program will enhance the existing recycling market and create opportunities for locally based businesses and social enterprises.
2. Handle packaging and printed materials in a way that assures their highest and best end use.
3. Refrain from incineration or landfilling of recyclable materials.
4. Demonstrate progress up the pollution prevention hierarchy.
5. Establish and achieve a measurable recovery target and timeline for each separate category of material included in the basket of goods.
6. Set recovery targets using the highest-performing collection systems as benchmarks (e.g., CDR system recovery rates).
7. Monitor program results and issue regular public reports that are available for public viewing online.

A Critical Test Case for EPR Now Underway in British Columbia

In May 2011, the British Columbia government officially designated packaging and printed paper as the next EPR program to be established. By November 2012, the producers of essentially anything sold in a package, along with the producers of printed paper (including newspapers, magazines, directories, and advertising circulars), are obligated by law to submit plans for taking back and recycling these materials.

Ontario Blue Box: Performance and Cost, 2009

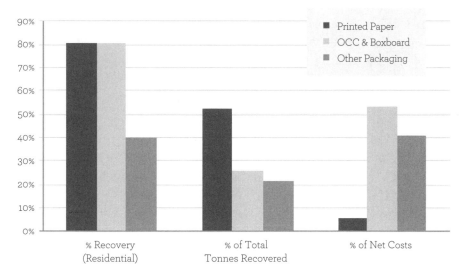

FIGURE 18.1. The performance and cost of the Ontario blue box program. *Source: Product Policy Institute*

To no one's surprise, these diverse industries formed an association (Multi-Material British Columbia [MMBC]) to develop the plan. Every indication is that MMBC is preparing to propose and then roll out a provincewide multimaterial curbside collection program on the same model as the one carried out in Ontario and proposed in Vermont and Minnesota. (The CDR system for beverage containers will remain in effect.)

The local context includes a strong risk that the plan will include a proposal to use incineration for some of the materials coming back under the plan. The Metro Vancouver region's recently adopted regional waste management plan includes a waste incinerator and the intent to seek provincial authorization to burn nonrecyclable products and packaging collected through EPR programs.[18]

But in the spring of 2012 there was an intriguing new development.

The *Vancouver Sun* reported in April[19] that the Canadian Newspaper Association is quitting MMBC. Newspaper Association CEO John Hinds complained that MMBC is putting together "an inefficient, expensive system that works for meat wrappers and yogurt containers but not for newspapers."

Hinds suggested that the newspaper industry is considering putting together its own collection system: "We have a very good one way delivery system right now. Maybe that has to become a two way delivery system." This statement is a positive indication: The newspaper industry is beginning to think in cradle-to-cradle terms.

For the other members of MMBC, however, keeping newspapers in the multimaterial system is critical to controlling costs and for meeting diversion targets. Figure 18.1, prepared by the Product Policy Institute from Stewardship Ontario, shows how exposed packaging will be without paper in the program. MMBC's chair admitted as much in the *Sun* article: "If you don't have newspapers, I don't think you have a viable blue box."

John Hinds and the Canadian Newspaper Association have, at the time of this writing, been quiet since the announcement, but a schism could stymie single-stream collection. Furthermore, separation of newsprint will highlight the low recycling performance of designed-for-disposal packaging that is so prevalent. Using printed paper as the model could permit establishing robust measures in place for setting targets, monitoring results, and preventing incineration for packaging. Such a development would put direct pressure on packaging that will reverberate through the industry.

Producer Responsibility, the Cornerstone of Zero Waste: Trends in the United States

By Bill Sheehan

BILL SHEEHAN

*Bill Sheehan has been as passionately opposed to landfilling as I have been to incineration. When we finally met in the late 1990s, the only solution that would satisfy both our passions was zero waste. Sheehan is a visionary policy advocate who has been at the forefront of two important sustainability movements in the United States. In 1995, he helped launch and lead the civic movement for zero waste as cofounder of the GrassRoots Recycling Network (GRRN), for which he was executive director for eight years.[1] It was in that capacity that Sheehan invited me to attend the California Resource Recovery Association's annual conference in San Francisco in 1999. The object of the exercise was for me to meet and videotape some of the leading theorists and practitioners of zero waste. The videotape that emerged—*Zero Waste: Idealistic Dream or Realistic Goal?[2]*—can be viewed online. In 2001 Sheehan and I published a booklet entitled* A Citizen's Agenda for Zero Waste: A United States/Canadian Perspective.[3]

In 2003 Sheehan cofounded the Product Policy Institute[4] to focus on public policies that address unsustainable production and consumption at the product- and packaging-design stage. Using a historical analysis of municipal waste management, he helped local governments organize product stewardship councils to work for state producer responsibility laws in California, New York, Texas, Minnesota, Vermont, Rhode Island, and Massachusetts. Sheehan is currently working to launch a national coalition of grassroots NGOs working on state producer responsibility policies. He holds a PhD in insect ecology from Cornell University and held a research position at the University of California at Berkeley.

The policy approach known as EPR has been called a prerequisite for and cornerstone of zero waste by state and local government officials in the United States, Canada, and Australia. The reason is that a key obstacle to achieving zero waste is that too many products are designed to be thrown away or contain toxic components. No amount of recycling such products will ever be sustainable. Moreover, when taxpayers pay to recycle or otherwise dispose of manufactured goods, they are providing an incentive for manufacturers to keep designing and selling disposable, unrepairable, overpackaged, and toxic junk. EPR allows local sanitation departments to put resources into development of infrastructure for managing the putrescible wastes that are producing methane in landfills, while compelling producers to ensure that infrastructure is in place to deal with their products and packaging.

EPR (sometimes called product stewardship and manufacturer responsibility) aims to correct that market distortion by making producers—the parties that design, make, and sell products—responsible for ensuring that systems are in place to responsibly recycle their products, at no cost to taxpayers, *before the products can be sold.* Government has a critical role to play by setting performance standards and outcomes in the public interest, allowing industry flexibility to innovate to achieve those outcomes, and finally ensuring transparency and accountability. Outcomes beyond recovery rates in the public (community) interest can include utilization rates (clean streams of recovered materials are used in making new products); reuse targets (which facilitate local economic development); fair compensation for public assets (for communities that elect to stop providing waste management services for manufactured discards); and incentives for competition among industry stewards (which drives improvements in product design).

The United States has lagged behind Canada and the European Union in implementing producer responsibility policies for manufactured discards. Now the producer responsibility approach to managing products and packaging is spreading like wildfire in the United States, with almost fifty state laws having been adopted since 2004. The issues and landscape have changed significantly since 2010, with the broadening of the scope of products targeted and with industry starting to push back.

Canada is leading the way in North America. In 2009 the Canadian Minister of the Environment set a goal of applying producer responsibility policies to virtually all manufactured discards by 2017.[5] The province of British Columbia established the first EPR framework regulation in 2004, has the most products covered, and pioneered the approach in which government sets performance standards—product category by product category—and then lets industry propose a plan to achieve the desired outcome.[6] Nova Scotia has also been a leader in EPR.

EPR moved down the west coast of North America over the past decade and has also spread eastward, challenging the unfettered autonomy of corporations and voluntary programs promoted by the US EPA and other stakeholders who advocate for "shared" responsibility.

Environmental NGOs have been an important force in bringing EPR to the United States. The Electronics TakeBack Coalition[7] worked with state-based organizations to pass many of the twenty-three state producer responsibility laws for discarded electronics. Other national product-focused NGOs work on EPR for beverage containers (Container Recycling Institute[8]) and mercury-containing products (Mercury Policy Project).[9]

Product Policy Institute (PPI) is the main national environmental NGO focusing on EPR as the cornerstone of zero waste. PPI was founded in 2003 when British Columbia was developing its framework producer responsibility regulation. PPI's research on how local government waste management programs have enabled the production of throwaway products resonated with local governments faced with unfunded mandates and budget crises. PPI organized local governments in product stewardship councils to work for statewide EPR legislation and ultimately framework legislation like British Columbia's.

In 2006 PPI helped start a council in California[10] based on the model of the existing Northwest Product Stewardship Council.[11] The California council became a model for others in New York,[12] Texas,[13] Vermont,[14] Minnesota, Connecticut, Rhode Island, and Massachusetts. PPI guided development of principles adopted by the councils, recently led development of a broader consensus document for advocates titled Product Stewardship and Extended Producer Responsibility,[15] and promoted the adoption of local EPR resolutions—currently more than 150 in six states.[16] The three major national associations of local elected officials—National League of Cities, National Association of Counties, and US Conference of Mayors—also adopted PPI's producer responsibility resolution.[17]

In 2010 PPI cofounded and organized Cradle2,[18] a national coalition of public interest organizations focused on source reduction and recycling of consumer products and packaging. Cradle2 is working to build the political power to have states adopt EPR policies for virtually all products and packaging in the waste stream. While member organizations are addressing a wide array of products, EPR for packaging is emerging as a top priority.

EPR is entering a high legislative phase in the United States, similar to the period between 1988 and 1992, when dozens of state recycling laws were passed (unfortunately, with no mention of producer responsibility). Currently, almost eighty producer responsibility laws covering nine categories of mostly products are on the books in thirty-two states—with the large majority adopted since 2004.[19] The hallmark of most of these laws is that they are product-specific, address mostly hazardous products, and are state-level laws. But current trends show the EPR movement moving into new territory.

Framework EPR Legislation. In March 2010 the State of Maine became the first state to move beyond product-specific legislation to pass a comprehensive, or "framework," EPR law.[20] The law does not name any products but rather sets criteria and creates a process for bringing in new products over time. The state environmental agency reports annually to the legislature on product stewardship programs and proposes legislation when new products are deemed appropriate for producer responsibility. Other states have also been active, informed by PPI's EPR legislative starter kit.[21]

EPR for Packaging. Most attention so far has been on hazardous products, but attention is starting to turn to high-volume packaging. All eyes are currently on British Columbia, where packaging and printed paper were brought under the province's EPR regulation in May 2011.[22] All producers of packaging and printed paper have eighteen months to come up with plans for how they will manage the materials at no cost to taxpayers, and another eighteen months to start implementing the plans. In the United States, Nestlé Waters is actively supporting EPR for packaging and working through a new organization, Recycling Reinvented, to pass state legislation. The Cradle2 coalition is engaging to steer policy in the public interest direction and to build political power in targeted states.

Federal EPR Activity. There is currently one bill in Congress to stop the global dumping of electronic waste—something only Congress can do.[23] Given extreme dysfunction in Congress, no one expects broader federal EPR legislation anytime soon.

A trend intertwined in the above developments is that industries potentially obligated by EPR laws have started to engage in a serious way—both positively and negatively. The Maine framework EPR bill passed the legislature unanimously with Maine Chamber of Commerce support.[24] In California a paint EPR bill had industry support and no opposition, while battery and lamp EPR bills had strong industry opposition.[25] In September 2011 associations for carpet, forest product, and toy producers formed an alliance of trade associations, the Product Management Alliance, to oppose EPR legislation across the United States.[26]

This is an exciting time for EPR in the United States. The question is no longer whether we will have producer responsibility for recycling products and associated packaging. The question now is how to make producer responsibility the foundation of a zero waste economy.

Together at Last:
Extended Producer Responsibility
and Total Recycling

By Daniel Knapp

Daniel Knapp is founder and director of Urban Ore in Berkeley, California (see chapter 15 for his extended bio). In this chapter, Knapp examines ways that two previously opposing factions of the zero waste community—those advocating total EPR to eliminate waste and those advocating total recycling to eliminate waste—can find common ground. Knapp adapted this chapter from material originally prepared for presentation to the Northern California Recycling Association *in 2011 and published in* NCRA News, *the organization's online newsletter.*

DANIEL KNAPP

EPR versus Total Recycling

Although I support and believe strongly in producer responsibility, I've argued strenuously for total recycling.

I'm for EPR, always have been. But the people who grabbed the EPR brand and ran with it about ten years ago developed a rhetoric that assumed recycling was in the way and had to be set aside for EPR to work. This rhetoric often resorted to sloganeering: Recycling was "so last century"; recycling "enables wasting." They said EPR, pursued correctly, made recycling outmoded and unnecessary, because products would simply go back to their makers. Unfortunately, this rhetoric put the believers and practitioners of total recycling on a collision course with the theorists of EPR.

However, one event showed me that EPR *as practiced* is much closer to total recycling than this rhetoric would suggest. This event was an all-day webinar sponsored by the California Product Stewardship Council (CPSC), and the nine or so speakers gave a grounded and factual picture of how EPR is actually working.

And how it is working is not like the EPR framework says it should be working. It needs recycling facilities to succeed at all.

The webinar focused on just one commodity type: batteries. The speakers were actually part of the battery reclamation supply chain in various parts of California. My big takeaway message from a day of listening was EPR ideas are being tested and refined in actual practice; reality is forcing EPR and total recycling back together.

They are complementary, not opposed. Both are necessary for attaining zero waste to landfill and incineration. Here are six reasons why.

Former Opponents Are Getting Together

Helen Spiegelman of the Product Policy Institute (PPI) and I have debated both sides of the EPR issue for years on the GreenYes listserv. Others often joined in. These discussions erupted spontaneously, usually stimulated by some event or other that we could use as a launching point. They could be contentious, but more recently Spiegelman and I have been e-mailing back and forth in tones of reconciliation and mutual respect. Spiegelman seems as dismayed as I am about some of the ways EPR is being implemented. Besides writing some kind words about Urban Ore as a reuse business, in one of her responses Spiegelman praised the Institute for Local Self-Reliance (ILSR) and Urban Ore for defending EPR from the "giants of garbage" who want to "highjack" EPR commodities to feed landfills and incinerators.

EPR Needs a New Generation of Total Recycling Facilities

Batteries are one commodity whose final disposal just about all EPR programs are designed to control.

One of the speakers on the webinar was Tedd Ward, who with Kevin Hendrick (the vice president of CPSC) manages the regional recycling and waste transfer station in Del Norte County, California. This is a rural county six hours by road from San Francisco, so it is distant from export markets. Nevertheless, the county has had great success in waste reduction, having exceeded the statewide 50 percent diversion mandate a couple of years ago.

Ward stated that the Del Norte transfer station is the largest source of EPR-collected batteries in Del Norte County, despite the fact that most if not all retailers in the county now have takeback systems. He presented a graphic that shows battery reclamation at very low levels until 2008, when they jumped tenfold. The increase roughly coincided with the opening of the "regulated materials area" at the new publicly owned transfer facility.

According to Ward, the transfer station accounted for 3,065 pounds in 2009, more than nine times the contribution of the next two collectors—both large stores that sell the batteries. The PPI's framework for EPR would lead one to expect the reverse. But that is not what is happening.

Incidentally, the in-store takeback program was funded by a grant from CalRecycles, the new agency that in 2009 replaced the California Integrated Waste Management Board. Since the county population is small, the retailer takeback program has achieved very high participation by businesses selling batteries, so it is a good case study of the emerging industry.

This finding verifies that a new generation of transfer stations is needed even with full EPR at the retailer level. The reason a new generation of transfer stations is needed is that the old ones built in the 1980s to replace closing landfills are wearing out and were usually not designed to fit today's best practices in reuse, recycling, and composting. Many lack a designated area where regulated materials may be collected, sorted, and shipped to producers or end users.

Materials Handled Similar Between Canadian EPR Program and Californian Zero Waste Transfer Stations with Regulated Materials Areas

In May of 2011, Monica Kosmak, a zero waste planner for Vancouver, summarized the regulated materials covered by the Vancouver, British Columbia, EPR program—a program that Bill Sheehan of PPI has called "the most comprehensive, industry-supported Extended Producer Responsibility approach in Canada." That list, below, is a fairly complete overview of the universe of EPR commodities circa 2011.

- Electronics
- Paint
- Pesticides, solvents, gasoline
- Used oil and empty oil containers
- Oil filters
- Tires
- Lead acid batteries
- Compact fluorescent bulbs, thermostats, and unused medications
- Antifreeze and empty antifreeze containers
- Small appliances

I believe that all of the above commodities are currently handled by and within the Del Norte transfer station regulated materials area. Some are dropped off at or near the purpose-built "household hazardous materials" building on the site, and some are salvaged from the wasting area tip floor and brought there for processing.

Customers pay a disposal service fee to enter the facility and may also pay additional commodity-specific fees in order to drop off certain regulated items. Other collections may be partly or fully supported by manufacturers or payments from the state. In contrast to what the PPI's framework requires, the county solid waste authority (rather than the manufacturers) is currently bearing the full cost of sortation, packaging, and shipping. So governments looking to save money by implementing EPR may find their costs increasing rather than decreasing.

My guess is that these days just about any transfer station or landfill that is open to the public will have a place or places where most of these commodity types can be handled.

Many EPR Battery Professionals Think Space, Labor, and Funding Needs Hold Back EPR's Further Development

Controlling the collection of targeted materials with mandates is just the first step in EPR. Next comes processing. In the EPR webinar I heard again and again from different speakers representing all levels of the battery supply chain that the space and funding to do the labor- and skill-intensive work of EPR processing is a big problem that must be solved. Rob D'Arcy, program manager for the Santa Clara Hazardous Waste Recycling and Disposal Program, said, "You have to be careful what you wish for."

He was referring to being inundated by regulated toxic and hazardous materials soon after the start-up of takeback programs. He said that currently "the takeback concept is burying local governments." He gets some of his funding for state-mandated universal waste handling from PG&E (a regional electric utility) and has been thinking of asking them to earmark 60 percent of any funding for *lighting* takebacks to be spent on "infrastructure"—space and labor to handle the fluorescent fixtures, tubes, and compact fluorescent lights (CFLs). He said his agency's ability to pay for processing costs for these and other materials is almost nil and that his impression was that "household hazardous waste programs are among the best-kept secrets of local governments."

Sortation requirements can be labor-intensive and costly. What kind of work must be done? Batteries, for example, which collectively are less than 1 percent of the discard supply, must be sorted quite carefully. Leaving aside the question of auto and other large batteries, there are at least five major types of smallish batteries from gadgets and appliances such as TVs, computers, and phones—alkaline rechargeable, alkaline nonrechargeable, lithium ion, nickel-cadmium (NiCad), and nickel–metal hydride (NiMH). Some battery types are dangerous and can explode if mishandled. NiCad and batteries larger than nine volts have to be taped and bagged. Sortation and prepping for shipment off-site is a major cost for collectors, and too often it is unpaid. At the Del Norte transfer station, for example, sorting and packaging batteries for transport now takes about three person-days per month. At the much bigger and more automated Recology transfer station south of San Francisco, sortation takes ten hours of hand labor every day. One of the webinar speakers reported that, at his facility, one person stands at a table all day every day doing nothing but sorting batteries. He said the work was "like shuckin' oysters."

Other commodities covered by EPR mandates have similar processing requirements. The webinar was full of real-world examples that confirm that lots of space as well as careful human handling is needed to make EPR work. So EPR done well creates lots of jobs. It is labor- and skill-intensive, just like reuse done well.

Although the space needs of EPR are large, EPR policy does not provide much in the way of specific guidance on what kinds of spaces are best. One representative from Energizer batteries laid out a vision for an industry-sponsored rollout of EPR battery takeback, and said any program they back will have to be multibrand, involve "all levels of the supply chain," be implemented in phases, and based on "trials" of different ways to handle the materials. He asked participants to "be patient with us" as the industry figures out what to do.

Urban Ore–Type Handling Systems
Can Model New Zero Waste Transfer Stations

Nearly all airports are owned and operated by an authority of some kind that builds and maintains the infrastructure while collecting rents and user fees to finance itself. This same model can be applied to transfer stations. The authority running the transfer can provide space to anywhere from one to several specialized materials processors and operators. It can charge rent for that space, and it can also charge a gate fee, an important source of day-to-day funding. EPR programs could finance the processors whether they are operating takeback programs in stores or in transfer stations.

Reconcile EPR and Recycling by Helping Recyclers Build Proper Zero Waste Station Networks

Do this, and we're on the march. Then we can take our ideas and designs for such transfer stations to a federal government still trying to find projects that they can finance to soak up some of the excess labor now sloshing around in the US economy. How about a rollout of zero waste transfer stations like Del Norte to replace the landfills and incinerators, and over time to replace the dirty material recovery facilities, too?

Zero waste programs like Vancouver's recommend reducing and reusing; capturing the organics; keeping recyclables out of landfills and incinerators; and providing for construction, renovation, and demolition recycling. They also foster a local closed-loop economy. All these worthy goals are best served by comprehensive twelve-market-category zero waste transfer stations, in addition to other strategies such as retailer takeback.

A review of EPR statements from the Sierra Club, ILSR, Clean Production Action, and GRRN shows that all are prepared to be somewhat flexible in determining when something is "true EPR" and when it is not. ILSR, for example, says, "EPR initiatives include product take-back programs, deposit-return systems, product fees and taxes, and minimum recycled-content laws." The US EPA says, "There is no 'one size fits all' EPR solution appropriate for all product systems."

Urban Ore has never been against EPR itself, just what we see as overly rigid interpretations and a tendency to label recyclers as "enablers" of wasting. That now seems to be behind us.

Together we can do great things.

The Economics of Zero Waste

By Jeffrey Morris

Economist Jeffrey Morris is principal of Sound Resource Management Group (SRMG), an economic and environmental research and consulting firm located in Olympia, Washington.¹ Morris has taught economics (including input-output analysis), econometrics, life cycle analysis and assessment, and ecological and social sustainability at the undergraduate and graduate level. He has also published in peer-reviewed journals on economics, life cycle analysis, and industrial ecology. He holds a PhD from the University of California at Berkeley.

JEFFREY MORRIS

Morris has been analyzing and writing on the economics and environmental impacts of municipal solid waste management systems for twenty-five years. During that time he has developed a solid waste utility rate-making model, as well as a methodology for interjurisdictional rate comparisons, both of which continue to be used by municipal public utilities and public works agencies. Morris also developed an econometric separated waste stream forecasting and evaluation model for Seattle Public Utilities (SPU) that SPU continues to use.

From 2006 to 2007 Morris led a national team of economists and life cycle experts to create a consumer environmental index (CEI) for the Washington State Department of Ecology. This was and still is the first successful indexing methodology using a consumption-based inventory for tracking the environmental footprint of consumer final demand over time. Recently, Morris has developed an outcomes-based life cycle assessment tool, MEBCalcTM, for measuring the environmental footprint of waste management systems.

I first met Morris on Long Island in the late 1980s. His company provided an economic analysis comparing the costs of an incineration project with the alternative of intensive recycling and composting. This allowed local decision makers to use a bottom-line analysis to side with citizens who wanted to defeat the incinerator for environmental and health reasons—and the project was defeated. Many years later Morris's company provided the very same service—with the same result—in Halifax, Nova Scotia. The end result of that victory was the successful program that Nova Scotia proceeded to pioneer.

The economics of zero waste is about financial and environmental costs that are avoided when discarded products and packaging materials are reused, recycled, or composted rather than buried or burned. It's about avoiding/ preventing discards in the first place by reducing consumption of goods and services that yield discards. It's about removing the subsidies for virgin raw materials extraction, as well as passing clean air, clean water, and public health regulations that ensure that virgin materials users pay the full environmental costs of extracting raw materials from ecosystems and processing them into materials used to manufacture new products and packaging. And it's also about jobs created when discarded products and packaging are used to manufacture new materials and products.

The major environmental benefit from reduce, reuse, recycle, and compost (three Rs + C) is that fewer raw materials need to be extracted from ecosystems and refined into materials suitable for manufacturing products and packaging. Thus, 1) with reduction measures less-raw-material- and energy-intensive goods and services are chosen by consumers, 2) reused objects displace newly manufactured items, 3) recycled materials displace newly extracted raw materials, and 4) compost replaces synthetic fertilizers and pesticides. These choices all reduce the "upstream" environmental footprint caused by our consumption habits and needs.

No less important is that the three Rs + C reduce the pollution that comes from the disposal of these same materials via landfilling and/or incineration. Typically, these releases are not fully offset when the energy generated from captured landfill gas or from combustion of discards in incinerators is compared with other means of generating energy. Sometimes these are compared with coal-burning facilities but more appropriate today would be to compare them with natural gas facilities. In the future these "next-in-line" energy sources will be trending toward solar or wind. In both current and future situations these displaced energy sources do not release as much pollution as is released when energy is generated from buried or burned discards. Hence, environmental footprints of landfills and incinerators, even after taking into account pollution offsets for energy generated from discards, still need to be avoided. And hence, environmental burdens of disposal that are avoided by the three Rs + C add to the environmental bottom-line benefits of diverting discarded products and materials to three Rs + C programs.

At the same time, it's also true that reuse, recycling, and composting generate their own environmental footprints from collection and processing. These systems are employed to make diversion of discards from disposal convenient, as well as to sort and transform collected discards into reusable consumer goods or materials useful for manufacturing recycled-content products and packaging. The environmental footprints of these collection and processing systems typically are small in comparison to the environmental impacts of products and packaging produced from virgin raw materials and the impacts of discards disposal. So the environmental impacts avoided by reuse, recycling, and composting are much larger than the impacts of collection and processing systems employed to divert discarded products and packaging materials from disposal.

In addition, successful diversion systems can lead to substantial downsizing in trash collection and transfer systems. This further enhances the environmental benefits of the three Rs + C. For example, figure 21.1, based on an econometric analysis by SRMG, details the steps taken by SPU to divert discards to recycling and composting and decrease disposal quantities. The graph shows average disposal quantity per garbage collection day for Seattle residential households (both single- and multifamily buildings).

As indicated on this graph, when Seattle first introduced pay-as-you-throw (PAYT) variable rates based on garbage can size for garbage collection in 1981, there was no detectable change in residential waste disposal. But as soon as Seattle introduced curbside recycling and composting collections in late 1988/ early 1989, residential waste disposal dropped dramatically. This indicates the importance of having recycling and composting collections available that are as convenient as (and less costly than) garbage collection. After initiation of PAYT but before introduction of curbside diversion programs, Seattle had become somewhat infamous for the "Seattle Stomp." That is, PAYT without convenient recycling caused residents to stomp their garbage into smaller garbage can sizes in order to reduce their garbage collection costs.

In addition to offering curbside recycling and composting collections, Seattle has used other regulatory and financial incentives over the years to motivate residents to further reduce their disposal quantities. Regulator initiatives have included bans on the collection of yard waste and mainstream recyclables in the garbage, augmentation of the list of materials that can be put in the curbside recycling container, universal food waste collection in yard waste containers, and enhancement of PAYT by charging garbage collection fees

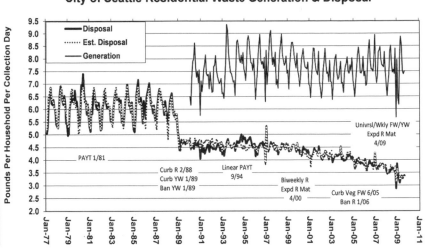

FIGURE 21.1. Timeline for discard collection rate trends in Seattle. *Graph courtesy of SRMG*

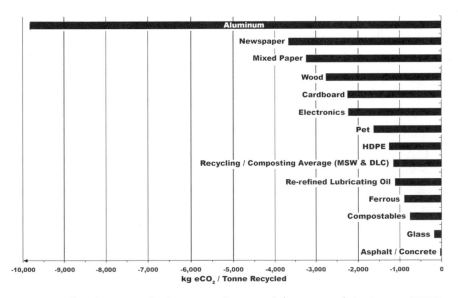

FIGURE 21.2. Greenhouse gas reductions per metric ton recycled or composted. *Graph courtesy of SRMG*

that increase directly in proportion to garbage container sizes (in other words, if a household doubles the size of the garbage container it uses, it pays double).

To illustrate one of the environmental benefits of increased diversion and reduced disposal of discards, figure 21.2, from a study by SRMG on the benefits of a zero waste plan for the Metro Vancouver region, shows the reduc-

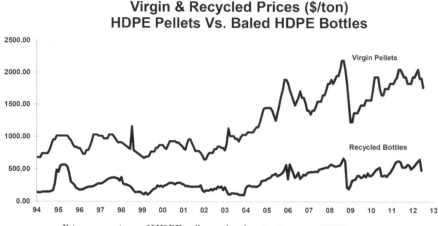

FIGURE 21.3. Price comparison of HDPE pellets to bottles. *Graph courtesy of SRMG*

tions in climate-changing greenhouse gas emissions when discarded materials and packaging are diverted through recycling and composting collection and processing systems rather than collected for burning and burying at disposal facilities. More details and studies on how to quantify the upstream raw material and downstream disposal environmental footprints avoided by the three Rs + C are available at SRMG's website (www.zerowaste.com). In addition to climate change, environmental footprint indicators quantified in these studies include human respiratory, toxicity, and carcinogenicity impacts, as well as acidification (acid rain), eutrophication, and ecosystem toxicity impacts.

Unfortunately, many of the environmental costs of production or disposal are not adequately reflected in the prices we pay for products and packaging. Furthermore, there are numerous subsidies and tax incentives for virgin raw materials and energy use (for example, see the information provided at Earth Track's website, www.earthtrack.net).[2]

What these externalized costs and virgin material subsidies mean for the financial economics of diversion is that prices for reusable products and recycled materials and packaging discards are kept lower than they otherwise would be. This makes it more difficult for diversion systems to cover their costs with the revenues generated from selling reusable and/or recycled discards to manufacturers of recycled-content products and packaging.

How this works is illustrated by figure 21.3, comparing US national market prices for high-density polyethylene (HDPE) plastic pellets produced from virgin raw materials against market prices in the US Pacific Northwest for baled, recycled HDPE bottles. As indicated in the graph, virgin pellets always

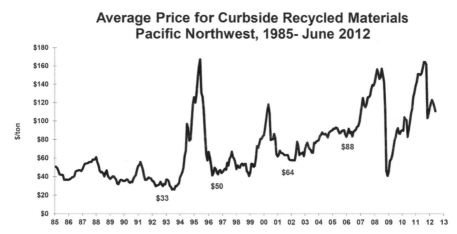

FIGURE 21.4. Price fluctuations for the sale price of curbside recycled material in the US Pacific Northwest from 1985 to June 2012. *Graph courtesy of SRMG*

yield a higher price than recycled bottles, due in part to the pellet buyers' being assured that the virgin pellets will exactly match desired specifications. The higher prices for virgin pellets are also due to recycled bottles' needing additional processing for conversion into recycled-content pellets.

Subsidies to virgin raw materials as well as externalized environmental costs of virgin raw materials extraction and refining keep virgin prices lower than they should be. Given the price cap that virgin materials place on recycled materials, virgin subsidies and externalized environmental costs depress the revenue available to recyclers to cover their costs.

Figure 21.4 illustrates another issue that recyclers encounter—the large fluctuations over time in market prices for recycled materials. The graph shows the average value per ton that US Pacific Northwest MRFs have been paid for curbside recyclables. The graph shows the weighted average price per ton for recycled newspaper, mixed paper, cardboard, glass containers, HDPE and PET food and beverage containers, aluminum cans, and small ferrous scrap (including steel-plated tin cans) collected via curbside recycling programs. This weighted average price per ton has ranged between $30 and $167 over the past twenty-five years.

Underlying causes of some of the price movements shown on the graph include the following:

- supply increases from new curbside recycling programs (late 1980s to early 1990s);
- general economic slowdown (late 1980s to early 1990s);

- manufacturing feedstock inventory buildup in anticipation of shipping container shortages during the 1991 Persian Gulf War;
- increases in recycled-content manufacturing capacity and demand, especially in the paper industry (early to mid-1990s);
- manufacturing feedstock inventory buildup in anticipation of continued price increases or supply shortages (1994–95);
- recession in Asian economies (late 1990s);
- unanticipated inventory shortages at domestic and foreign paper mills (especially in China, Mexico, and Indonesia), Y2K fears, and substantial recovery for sales of many grades of paper and paperboard (1999 and early 2000);
- generalized exuberance, dare we say "irrational," in asset and commodity markets (mid-2006 through mid-2008);
- the crash of financial markets in late 2008;
- gradual recovery back to more normal cyclical trends beginning in early 2009.

The three prices under the graph's cyclical trend line show average values for curbside recycled materials during months at the bottom of the three major recycled material price cycles over the period prior to the 2008 financial markets meltdown. A striking result is that bottom-end prices have moved up substantially from an average level of $33 per ton during the 1991–93 downturn to $64 in the 2001–02 downturn. In constant (2003) dollars bottom-end prices have increased 43 percent from $46 per ton (1991–93) to $66 (2001–02). In addition, over the cyclical flattening in 2005–06 and the pause in price increases in 2010, prices stayed above $80, up again from the cycle bottom average of $66 in 2001–02.

So the downside price risk to recyclers has been significantly reduced over the past twenty-five years, except that recycling markets will always be vulnerable to systemwide economic shocks such as occurred in late 2008. Improved resiliency against cyclical downturns is primarily due to upward trends in market prices for recycled materials such as mixed paper and plastic bottles.

At the disposal end of the virgin products and packaging materials life cycle there are also externalized costs and market constraints that operate to depress the financial economics of reuse, recycling, and composting. For example, incineration facilities often require guaranteed disposal amounts from the communities that use them for disposal. These guarantees often take the form of "put-or-pay" agreements whereby the facility users pay a fixed annual fee that covers disposal of a stated number of tons. If the community generates lower discard quantities that fee is not lowered, but if the community generates greater amounts additional fees are collected. This places a constraint on the ability of communities to capture avoided disposal costs to help pay for their recycling programs, because their disposal costs do not decline when discard volumes go down as a result of successful diversion programs.

In terms of the jobs benefits from recycling there have been two recent studies that lay out in some detail the jobs that would flow from increased diversion. *More Jobs, Less Pollution: Growing the Recycling Economy in the U.S.* was sponsored by the BlueGreen Alliance, GAIA, NRDC, the Teamsters and SEIU unions, and Recycling Works! It covers the expected employment and environmental benefits of substantially increasing diversion of discards from disposal.[3] The other report, *Returning to Work: Understanding the Domestic Jobs Impacts from Different Methods of Recycling Beverage Containers*, was sponsored by the Container Recycling Institute and details the job increases from beverage container deposit-return systems that substantially increase recycling rates for beverage containers.[4]

In summary, the environmental and employment benefits of the three Rs + C compared against burying or burning with energy generation are substantial. It's the hidden and not-so-hidden subsidies and tax breaks for upstream natural resource exploitation, and the continuing free ride for pollution of our air and water, that often prevent progress toward a zero waste society. These constraints make it difficult to earn enough revenue from selling diverted materials and avoiding disposal costs to cover the costs of collecting, processing, and recycling the valuable resources from our discarded products and packaging materials. Even in the face of these challenges, many communities throughout the world are making dynamic advances in their three Rs + C programs and taking long strides down the road to zero waste.

Businesses Are Leading the Way to Zero Waste

By Gary Liss

Liss holds a BS in environmental engineering and masters in public administration and currently is a zero waste consultant in his own firm. He is a leading advocate of zero waste in the United States and has helped more communities develop zero waste plans than anyone else in the United States. Liss was a founder and past president of the National Recycling Coalition and was recently elected president of the new US Zero Waste Business Council. He is a founder and board member of the Recycling Organizations of North America (RONA), the GrassRoots Recycling Network (GRRN), and one of the leaders of the Zero Waste International Alliance (ZWIA) and the Zero Waste Brain Trust. In 2008 he was elected to the council in his hometown of Loomis, California, and he was mayor in 2010.

GARY LISS

I have worked with Liss since the 1990s. Liss loves to network people and resources and is always helping to share information that is incredibly valuable. Liss was one of the experts I interviewed for the video Zero Waste: Idealistic Dream or Realistic Goal?[1] *I was immediately impressed with his machine-gun-like delivery, his bubbling enthusiasm, his enormous energy, and his huge grasp of the waste issue. His enthusiasm has proved contagious with everyone he meets, whether they be environmental activists, decision makers, or businesspeople. In this essay Liss highlights the importance of identifying and recognizing zero waste businesses for the success of zero waste.*

Early Days

In 1997, when I was executive director of the California Resource Recovery Association (CRRA), this organization held the first zero waste conference in the United States in Monterey, California, under the leadership of board member Steve Suess and conference chair Robin Salsburg. The conference closing speaker was Jerry Brown (now the governor of California), who talked about the zen of zero waste. Over one thousand people attended, and the event was a major success.

At this conference, CRRA adopted the Agenda for the New Millennium that had been drafted by Tedd Ward (the zero waste coordinator for Del Norte County) that called for zero waste, ending welfare for wasting, and creating jobs from better product design and productive use of discards. A new technical council was formed under CRRA, the Global Recycling Council (GRC), to make sure the Agenda for the New Millennium would be pursued by CRRA after the conference.

Building on Success

In the keynote address I gave at the 1997 annual conference of the National Recycling Coalition, I stated that zero waste is the next step in the American success story called recycling. Every day, more than 100 million citizens do the right thing—they recycle. Now it is time to set our sights higher and start planning for the end to wasting resources and our reliance on landfills, incinerators, and other waste facilities. Zero waste is a policy, a path, a direction, a target; it's a process, a way of thinking, a vision.

A New Planning Approach for the Twenty-First Century

Zero waste represents a new planning approach for the twenty-first century. The American economic system stands for individual freedom, entrepreneurship, and free-market capitalism. Zero waste adds to that system

- the principles of conserving resources,
- minimizing pollution,
- maximizing employment opportunities,
- maximizing efficiency, and
- providing the greatest degree of local economic self-reliance.

The key measure of whether zero waste is attainable or not is the fact that some businesses are already achieving close to zero waste, diverting over 90 percent of their discarded materials from landfills and incinerators.

- 97 percent diversion: Mad River Brewing Company in Northern California

- 95 percent diversion: Zanker Disposal and Recycling in San Jose, California
- 97 percent diversion: Hewlett-Packard in Roseville, California
- 95 percent recycling rates at office buildings in the EPA Green Buildings program

From the very outset of promoting zero waste in the United States, we built upon real successes reported by businesses to demonstrate that our vision was realistic and possible.

In 1998 at the Albuquerque, New Mexico, National Recycling Congress, I organized the first official session on the program on zero waste businesses. The session was standing room only, attracting one of the largest crowds at the conference, and afterward received the highest rating of any conference session that year. People were amazed to hear that zero waste was not just an idea, but that it was actually happening.

Zero Waste Business Principles

In 2001, Bill Sheehan, executive director of GRRN, obtained a grant from the Giles W. and Elise G. Mead Foundation to research zero waste businesses and develop a certification process. I did the research and was encouraged through that to develop zero waste business principles to guide businesses in a positive direction, without trying to develop a full certification program. GRRN adopted zero waste business principles developed through this project in January 2004.[2] GRRN asked the new Zero Waste International Alliance (ZWIA) to adopt similar zero waste business principles. ZWIA adopted them (slightly revised from GRRN's after much input over the ZWIA listserv) in March 2005.[3]

Since that time, at CRRA conferences; NRC conferences; zero waste business conferences organized by Earth Resource Foundation;[4] and other national, state, and local events, we would get representatives from zero waste businesses to speak and tell their stories.

Zero Waste Businesses in Japan

In 2006, at the CRRA conference in San Jose, California, entrepreneur Gunter Pauli spoke about his Zero Emissions Research Initiative (ZERI). Over dinner, he highlighted how 2,800 businesses in Japan had adopted zero waste as a goal, and that 99 percent had reported that they had achieved zero waste to landfill, and 34 percent had achieved zero waste to incinerators. He said that ZERI had been established by a grant from the Japanese government to research how to get to zero waste to help meet the goals of the Kyoto Accords. The Japanese minister of the environment had asked Japanese companies to adopt zero waste and zero emissions as a goal, and asked ZERI to help train businesses on how to get there. Upon reflection on that information, we realized that many of the zero waste business leaders in the United States that we had heard about were Japanese corporations that had embraced this goal:

Anheuser-Busch, Fairfield, California
Apple, Elk Grove, California
Epson, Oregon
Fetzer Vineyards
Frankie's Bohemian Cafe, San Francisco
Greens Restaurant, San Francisco
Hewlett-Packard, Roseville, California
Honda
Mad River Brewing Company
New Belgium Brewing
Pillsbury
Playa Vista, Los Angeles, California
Ricoh Electronics
San Diego Zoo Safari Park
Scoma's Restaurant, San Francisco
Subaru
Toyota
Vons-Safeway
Xerox Corporation

As we have worked around the country to develop zero waste community plans, we have found that there is often at least one or more leading businesses in recycling in each community; sometimes we have found that they are diverting over 90 percent of their discards from landfills and incinerators. When we cite businesses elsewhere that have been doing that, and highlight the local businesses, it becomes a very powerful way for communities to embrace the idea of zero waste for themselves.

All Zero Waste Businesses Save Money

We have found from all these zero waste business presentations that all zero waste businesses have saved money. They save the most money from reducing and eliminating wasteful practices, they saved the next most from setting up reuse systems like returnable shipping containers, and they also save from recycling and composting if their community has set up programs and garbage rates to provide appropriate incentives for waste reduction. Businesses also have increased their efficiency; reduced their liability from having to pay their share of cleanup costs from failed landfills; reduced their greenhouse gases and other emissions; and obtained a positive image with their customers, their community, and their employees—which benefits the business significantly.

Since the downturn in the world economy, there has been an increasing interest among businesses about the concept of zero waste. Businesses are finding that it's harder to generate new revenue than it is to reduce their costs, so they start looking into what they can do more efficiently. They read stories that have been appearing in the mainstream media since 2001 about zero waste businesses[5-10] and begin looking into those. When they find that this is a money-saving proposition, they embrace it and pursue it as part of their sustainability and climate change initiatives. In the last two years, more and more businesses are asking for guidelines on what zero waste means, and they ask if anyone is certifying businesses that achieve zero waste. Other companies are claiming to be zero waste—but oftentimes are only zero waste to landfill and burn much of their discarded materials.

Fighting for the ZWIA Definition of Zero Waste

In response to that situation, ZWIA and GRRN have begun stressing the need to fight for the brand of what zero waste means. The worldwide environmental and recycling communities agree that the only peer-reviewed, internationally accepted definition of zero waste is that adopted by ZWIA.[11] The combination of the ZWIA definition of zero waste and Zero Waste Business Principles and Global Principles for Zero Waste Communities are the brand we all are fighting for. The ZWIA zero waste definition basically says that all discarded materials are resources; resources should not be burned or buried; and our goal is zero air, water, and land emissions.

US Zero Waste Business Council

Together with that effort, the Earth Resource Foundation has helped form the US Zero Waste Business Council (USZWBC) to develop third-party certi-fication of zero waste businesses. The USZWBC has now been incorporated and organized a major National Zero Waste Business Conference in June 2012 in Costa Mesa, California. The USZWBC will be following the lead of the US Green Building Council in how it structures its program, and has begun working on a certification program to meet the demand for this service requested by businesses.

Meanwhile, in January 2012 ZWIA adopted a program to recognize zero waste businesses through national affiliates. As of this writing, ZWIA has approved four countries as national affiliates for this program: Australia, Brazil, Canada, and the United States. In addition, the USZWBC was formed to develop a third-party zero waste business certification program modeled after the US Green Building Council.

The Zero Waste International Alliance (ZWIA): A Chronology

By Richard Anthony

RICHARD ANTHONY

Zero waste advocate Richard Anthony is a founder and board member of the California Resource Recovery Association (CRRA),[1] the GrassRoots Recycling Network (GRRN),[2] the Zero Waste International Alliance (ZWIA),[3] and Zero Waste San Diego[4] and director of the consulting firm RicAnthony Associates. He holds a BS and MA in political science and is an adjunct professor for zero waste culture and planning at Irvine Valley College. I have worked with Anthony since the late 1990s. We first met each other in the late 1980s, but at the time we were on opposite sides. He was working to get a refuse-derived fuel (RDF) facility built in San Diego County, California, and I was helping citizens oppose it. We met again in 1999, and this time we found ourselves on the same side: zero waste. Anthony was one of the experts I interviewed for the videotape Zero Waste: Idealistic Dream or Realistic Goal?[5] *In this essay he gives the chronology of efforts to communicate the zero waste message to activists and decision makers around the world.*

In 2002, I found myself on the scientific committee for a series of resource conferences (called the R-series) organized by the Swiss research organization EMPA (the Swiss Federal Laboratories for Materials Science and Technology).[6] When I was reviewing papers submitted for an upcoming conference to be held in Geneva (R-02), I found far too many devoted to incineration and not enough tackling the issue at the front-end. The director asked me what I would like to see instead, and I said a workshop devoted to zero waste. I told him that if EMPA could come up with accommodation and waived registration fees I would put a team of experts together and find a way to get them to Geneva. To my delight the director agreed, and I then set out to get a team together to talk about all the different aspects of zero waste. My team (I should say my *dream* team) consisted of Bill Sheehan, an expert on EPR; Jeffrey Morris, an expert on cradle-to-cradle and cost-benefit analysis; Daniel Knapp, president of Urban Ore, a for-profit reuse-resale enterprise; Paul Connett, chemistry professor and noted international advocate for zero waste and sustainability; Bill Worrell, manager of the San Luis Obispo Solid Waste Authority; Joan Edwards, the former head of New York City and Los Angeles recycling programs; and myself.

I then had to go about getting the funds to get them all to Switzerland. When I approached Bill Sheehan, the director of the GRRN, he said that he would only help if this action was going to be more than an academic exercise. He insisted that we find a way to help community groups working on waste at the grassroots level.

With the help of some activist groups in the United Kingdom—Greenpeace UK and Communities Against Toxics (CATS[7])—we managed to set up a forum at Sussex University, in Brighton, to take place two days after the R-02 meeting in Geneva.

Many attendees were a little surprised to hear the details of the zero waste movement as presented by our team. This approach was quite novel to them. They were still largely bogged down in the safety and efficiency of landfills and incinerators. If they saw a role for recycling it was very much in terms of "integrated waste management" running alongside its big industrial partners.

After the Geneva conference most of the team traveled to the United Kingdom for the Brighton conference. This meeting attracted many grassroots activists fighting incinerators and landfills and others promoting recycling. Moreover, once the news got around, it attracted many decision makers from towns and counties across the United Kingdom struggling with the waste issue. Over one hundred people attended a very successful conference.

The next day in a nearby hotel, activists, most of our team of experts, and several others from the United Kingdom held a meeting chaired by Robin Murray, a professor at the London School of Economics who has written two books on zero waste.[8, 9] Together we helped to form a zero waste coalition for the United Kingdom and also set the wheels in motion for formation of ZWIA. This meeting was what we later called retrospectively the *first zero waste dialogue*.

In 2003, the *second zero waste dialogue* occurred at a conference held by the Global Alliance for Incineration Alternatives (GAIA) in Penang, Malaysia.

There was a moment when we heard the news that the United States had started bombing Iraq that we stopped our discussions. After a prolonged and emotional pause people at the meeting expressed the need for a positive approach to stopping all pollution and wars over resources. Most of us agreed that zero waste was a really good place to start. This really was the first truly *international* dialogue (the Sussex University meeting had been more of an Anglo-American affair).

At the first official meeting of ZWIA's Zero Waste International Planning Group, held later that year in Wales, we tried to put an international group together. We were able to agree on the principle and goals of ZWIA, but how to make the group representative and still registered as a charity became an issue still unresolved today. After this meeting, some of us went on to speak at the Welsh Recycling Conference in Llandudno, Wales—and held dialogues in Cardiff, Wales; Cork, Ireland; Doncaster, England; and Edinburgh, Scotland.

A *third dialogue* took place in San Francisco in 2004. This dialogue tied in with the recycling tours being organized by the US National Recycling Coalition and gave attendees a chance to see the San Francisco approach to zero waste. Nearly three hundred representatives from around the world came to the event to report on their progress and share current strategies and available technologies. An agreement was reached that ZWIA should continue planning for more international meetings and create a website and a listserv.

A *fourth dialogue* was attempted at the R-07 Conference in Davos, Switzerland, in 2007. A handful of us (a group from Lucca, Italy; five Americans; and two Romanians) had a great face-to-face dialogue, which was broadcast to the world via the web. We had people logged in from England, Hungary, the Philippines, the United States, and Australia. At the end of our dialogue there was a motion and a second to approve the following resolution:

1. support Revolve (a reuse operation in Canberra, Australia, being squeezed out of business) and other local struggles for zero waste;
2. support efforts to keep organics out of landfill;
3. support efforts to pressure manufacturers and retailers to take responsibility for their products and packaging;
4. support efforts to persuade local authorities to adopt source separation; and
5. support efforts to persuade local universities to set up resource management training.

This motion was passed by those who had joined us on the web and then later ratified via the ZWIA listserv.

The *fifth dialogue* was held in Naples, Italy, in February of 2009. The choice of Naples to host this event came as a direct result of the waste catastrophes Naples experienced in 2008, which led it to be called "the City of Fires." The international delegation brought a global focus to the waste problem in this area of Campania. More than one hundred Italian activists (the majority from Naples itself) came to listen to presentations from around the world, including the United States (San Francisco and Washington, DC), the Philippines, Canada, Bulgaria, England, Scotland, and Catalonia. The aim of the conference was to share knowledge and skills, to create positive action and support for communities fighting incinerators and landfills and promoting zero waste in the Campania area as well as the rest of Italy.

Delegates also heard directly from Italian community representatives on their experiences of waste problems in the surrounding areas. The international delegation also took part in a demonstration and march through the city center of Naples. At the end of the march, Italian and international delegates met with the local authorities to demand a rethink on their plan to build four massive incinerators to solve the Naples crisis.

After intensive discussions between the Italian attendees and international delegation, the conference adopted a Naples Manifesto on zero waste and the Global Principles for Zero Waste Communities.

It was with enormous joy that ZWIA heard, in 2011, that the newly elected mayor and vice mayor of Naples adopted a zero waste strategy for the city. This was an extremely courageous step on their part because it involved taking on the central Italian government, the regional government, and the Camorra—all of which are intent on building more landfills and incinerators in the area.

In 2009, the *sixth dialogue* was held in Puerto Princesa, the Philippines. This conference focused on the political issues of sustainability in both developed and developing economies. In particular, conference attendees examined the issue of waste management in terms of global warming. ZWIA speakers came from Europe, the United States, Africa, Asia, and India, and represented the latest thinking on sustainability without pollution during this world financial and environmental crisis. Present at the meeting were a number of people planning to attend the upcoming global climate conference in Copenhagen, and much of the discussion was on refining the message that these delegates would take to this conference. The ZWIA board adopted the slogan "The soil is the solution" as a way of tying together the triple goals of zero waste, zero warming, and zero toxics and the need to address these goals in the global efforts to "cool the climate" with a special emphasis on composting. The engagement of the Global South in articulating the issue of climate change and management of world resources was a critical focus.

In 2010, the *seventh international dialogue* on zero waste was held in Florianópolis, Santa Catarina, Brazil. This meeting brought together several hundred educators and students to dialogue about the new Brazilian resource management law and zero waste.

The *eighth dialogue* took place in 2013 in San Francisco, California. Key issues for ZWIA today include zero waste branding for corporations and government facilities, adopting zero waste plans at the community and business level, and compostable organics out of landfill (known as COOL).

ZWIA continues to grow as more and more people are realizing the need to shift from waste management to resource management as part of the transition toward a sustainable future. There can be little doubt that a sustainable society must be a zero waste society.

24

Response to the Ten Different Views of Zero Waste

There is a certain tension between the perspectives of Neil Seldman, Daniel Knapp, Mary Lou Van Deventer, and Buddy Boyd on the one hand and Bill Sheehan on the other. Seldman et al. want to maximize the use of discarded materials to provide employment and small business opportunities, especially in deprived communities. Sheehan would rather that much of this material never end up in the community by persuading manufacturers to take back their packaging and products. Sheehan appears to leave only composting for the community when he writes

> EPR advocates propose that community resources be invested in collecting and composting organic discards—food and yard trimmings, which produce methane in landfills—rather than in subsidizing management of manufactured discards.

That statement appears to leave little room for the community to reuse, repair, and recycle the materials that currently can be reused, repaired, and recycled—and provide jobs in the process.

I think that both sides would agree that the materials and objects that cannot be reused, recycled, or composted by the community should either be taken back by manufacturers or not produced in the first place. And we all agree that all our programs—whether EPR-oriented or community-oriented—are threatened by what Helen Spiegelman refers to as the municipal industrial complex (in other words, the waste industry and their friends in local government).

In chapter 20, Knapp offers a way to combine Sheehan's EPR vision with the practicalities of administering an EPR program by using current reuse (or "total recycling") operations—especially for hazardous and other hard-to-recycle materials. He argues that reuse operations and especially ecoparks are well set up to handle these materials if industries that have to meet EPR targets are prepared to pay a fee to these operations for doing so. Eric Lombardi (chapter 16) provides an example of where this is happening—the Eco-Cycle operation in Boulder, Colorado, called the CHaRM (the Center for Hard-to-Recycle Materials) facility.

Pressure for Better Industrial Design to Eliminate Waste

There are also different perspectives among zero waste proponents on where the pressure on industrial design should come from. Should it come from the zero waste research center in the residual separation and research facility (Step 8 in chapter 2)? Or should it come from state and federal legislation directed at the very front of the process, as Sheehan is advocating through the Product Policy Institute? I suspect that like most things in this business we will find ourselves working at both ends of the system at the same time.

Fighting Incinerator Proposals

From the perspective of fighting incinerator proposals, I think that Seldman et al.'s ability to provide alternative practical and cost-effective solutions that can be instigated immediately at the local level offers more hope to communities than Sheehan's more long-term theoretical perspective of effecting fundamental change at the front-end through legislation, as much as I would like to see this happen. We might well see this matter played out in Vancouver, British Columbia, which has some of the most advanced EPR activities in North America but is also confronted with major incinerator proposals, as discussed by both Boyd and Spiegelman.

If Sheehan's legislative efforts on EPR could also be combined with legislation to secure a moratorium on building incinerators at the state level, this would be a way to relieve the pressure on both decision makers and activists at the local level. By avoiding the waste of time, resources, and political energies involved in promoting and opposing incinerators, all our energies could be put into rational solutions that both citizens and local politicians could live with.

As Seldman points out in his essay, after discussing the defeat of over three hundred incinerators in the United States between 1985 and 1995, "It was with a huge sigh of relief that we at ILSR were able to refocus on our key mission—building local economies through the best use of local capital, labor, and raw materials. We could again focus on creating jobs and new enterprises from discarded materials." That's a sentiment I can identify with strongly.

Propagating the Message

Another issue that emerges from these discussions is how the zero waste message is propagated. It has been largely the passion and energy generated by opposition to incinerators that has helped to drive many communities toward recycling and zero waste, especially in Italy. However, a huge challenge remains as to how to generate the same enthusiasm for the zero waste strategy when there is no perceived immediate threat to the community or the economy. Can we get the same passion for sustainability? Can we get the same public support for the kind of measures at the front-end that Sheehan outlines? Are rational appeals for the planet's future sufficient to generate the same kind of pressure on political leaders to make them support zero waste? Ultimately, I believe it will be the visible benefits to the local economy in the forms of new jobs and new enterprises coupled with a regeneration of community spirit that will win the day for zero waste and sustainability. A visit to Urban Ore in Berkeley, Effecorta in Capannori, or restaurants, farms, and vineyards supporting San Francisco's composting program will speak for itself.

Can we anticipate more leadership coming from inside industry on these matters? How many more Ray Andersons, who made his Interface carpet company a leader in cradle-to-cradle manufacturing, are there out there? Gary Liss has persuaded us that there are quite a few. Most of the bad waste news from industry has come from the packaging industry. Most of the good news has come from the industries that make durable products. Ultimately, the zero waste message will prevail because the economics are right. As Liss reports, every business that undertakes a zero waste program saves money. Morris further elaborates on the economic advantages of the zero waste approach, especially when the environmental costs are internalized and not externalized. Sheehan's EPR message is one of the formal ways that such internalization may be effected.

Using the Internet

One of the things that all of the zero waste leaders featured in this text have seen change over the many years of their involvement in rational resource management and the zero waste movement is how their ability to propagate these ideas around the world has greatly increased with the advent of the Internet. Networking over vast distances has become very easy. This has proved most telling with Anthony's activities. While the international dialogues and conferences have provided invaluable face-to-face contact among some of the key leaders in the movement, they are very costly to organize. However, discussions on the Internet have taken the dialogue much further and at much lower cost. Today, many more people are talking about and doing what they can to further the zero waste message around the world. That zero waste message has both a community and industrial component.

The Ultimate Motive

When the newly formed US Zero Waste Business Council met recently in California, those in attendance came from many businesses moving toward zero waste and others that were interested in doing so. The accounts from representatives from Ricoh Electronics (copying machines), Albertsons (food store chain), MillerCoors (brewers), UPS (delivery services), Toyota (automobiles), and others were very inspiring. There were two messages that came across loud and clear. First, speakers underlined once again Liss's message that as they have reduced waste they have saved money. Second, to achieve maximum success they had to involve everyone in the company—from the top administrators down to the workers on the production floor. Not only has this made for very efficient programs, but it has also proved both empowering and psychologically uplifting for everyone concerned.

But the last slide presented by the representative from Albertsons, a national food chain headquartered in California, said it all. The representative explained that he had five reasons why he was putting so much time and energy into this effort. The slide showed his wife and four children. Nearly all of us who work on this issue are very conscious of the need to leave something behind for the next generation and many after that.

25

Conclusion:
The Fightback for the Future

Throughout this book, we've talked about the zero waste society as a stepping-stone toward a sustainable society, and we've seen that waste landfills and incinerators provide the visible evidence that our throwaway society simply cannot be sustained on a finite planet. We've also learned that zero waste begins with source separation—the ten fingers on everyone's hands. Then it moves through the very familiar territory of the three Rs of community responsibility—reduce, reuse, and recycle. We have stressed the need to combine community responsibility at the back-end of the problem with industrial responsibility at the front-end. Hitherto, much more attention has been paid to community responsibility. To focus more attention on industrial responsibility we have introduced a fourth R, which underlines industry's responsibility. This fourth R is redesign.

But one overriding reality is that everyone—from mom-and-pop recyclers to the brightest minds in our universities—needs to be involved to make zero waste a reality.

Nature Makes No Waste

Nature makes no waste; it is a human invention. Now we humans have to copy nature and reinvent an industrial world where we do not waste. That requires feedback mechanisms—at which nature excels. In the ten steps to zero waste presented in chapter 2 the crucial step is Step 8, which involves the need to build residual separation and zero waste research facilities in front of interim landfills. It is here that the feedback mechanisms must begin. The message that the community needs to send to industry is very simple and stark: "If we can't reuse it, recycle it, or compost it, you shouldn't be making it. We need better industrial design for the twenty-first century."

Some Corporations Leading the Way

Fortunately, a number of more progressive corporations have demonstrated that they can get very close to zero waste, and the even better news is that they are saving money doing it. For industry it is clear that solid waste is the visible face of inefficiency. However, the leadership is coming from the corporations making durable goods like automobiles, televisions, and copying machines. The companies more resistant to this idea are not surprisingly the companies making disposable goods, particularly those producing throwaway packaging. We need to find a Ray Anderson in the packaging industry. The most inspiring example that I have seen is the Beer Store in Ontario, Canada, which has been using refillable glass beer bottles for over sixty years, and saving money and creating many jobs doing it.

Community Responsibility

As far as community responsibility is concerned it is just that. I have stressed several times that countries don't recycle; communities recycle. Local political leadership is critical. A key feature of that leadership is the willingness of political leaders to work with citizens and activists. The time has long passed when politicians can solve their waste problems with highly paid "waste experts" promoting "magic machines." Magic machines won't save us: We need better organization, better education, and better industrial design. To succeed with these we need to draw on the expertise we already have in our own communities.

In the Global South, We Need to Respect Ragpickers

In the Global South the willingness of ragpickers (coupled with their desperation) to do their dangerous jobs of sorting through mixed waste needs to be respected. But they need to be given access to discarded resources *before* they are mixed together, the right to organize themselves into cooperatives and unions, and the opportunity to get a greater share in the profits made in their reuse. The examples from India and Latin America are very encouraging.

Instead of governments frittering millions away on multinational companies to build megalandfills and incinerators, a fraction of this money could be better spent providing the buildings and infrastructure to protect the health and welfare of the ragpickers. And more than that, they need to be given hope in the form of a career structure that would allow them to progress from scavenging for recycling, through to reuse and repair and remanufacture, and eventually operating and owning their own businesses. Eventually, their children may even play pivotal roles in the zero waste research centers that work on better industrial design and linking up zero waste with other aspects of sustainability.

The saddest example of where government has ignored the benefits provided by ragpickers—and even undermined their very existence—is in Cairo, Egypt. Ironically, the Zabbaleen in Cairo have shown us a model of how a community can integrate collection, separation , recycling, and remanufacture—albeit at a somewhat subsistence level.

In the Global North, We Need to Involve Our Farmers

In the Global North, our farmers are key players in an important step that dramatically increases diversion rates: composting. We need farmers involved in the whole process of moving from source-separated organic discards through efficient composting operations (or anaerobic digestion systems). It is key that when designing zero waste programs decision makers involve farmers early so that the compost produced can be used on their farmland with confidence. The farmers in turn need to be linked up with restaurant owners and community organizers who can ensure the quality of the separation programs. Farmers' markets are in vogue in many cities. It would be a good first step to involve these farmers in educating the shoppers they serve to encourage them to comply carefully with separation steps so that they, too, can be confident in the organic food that they are buying—grown with their own organic discards.

Getting Restaurants, Hotels, and Universities Engaged

San Francisco has also shown that in addition to involving farmers we need to involve restaurant and hotel owners who generate the bulk of the organic discards in large cities. Not only does San Francisco help to educate the kitchen staff in what materials do and do not get put into the clean organic containers, but they also provide an economic incentive to do so. Containers of clean organic materials are taken away for 25 percent less than mixed waste. In Italy, the Hotel Association of Capri has taken the lead in promoting zero waste.

If farmers and restaurant owners are needed to get us off to a great start with clean composting, then we need our academics to finish off the process. We need our professors and students involved in zero waste research. We need this not only to improve industrial design but also to link zero waste with other aspects of sustainability. It is going to be very hard to reach sustainability, but the thought of trying to do this without our universities playing a major role is unthinkable and a classic case of wasting resources. Zero waste research centers could provide our professors and students with real-life laboratories to start the paradigm shift from a throwaway society to a sustainable society.

Creating Jobs and Community

The job-creating potential of recycling, composting, and particularly reuse (low volume/high value) operations is enormous. They provide at least ten jobs for every job in landfilling or incineration. At the end of the day it is the real-life experience of job creation that is going to give the zero waste strategy its wings.

Reuse and repair centers offer even more than jobs. They offer the opportunity to recreate the village within the city. People enjoy looking for a bargain and in the process enjoy the people that they meet. These places help fight the anonymity of living in big cities. If our ultimate battle is against overconsumption then we have to offer something in its place. The only thing that I feel we can offer to replace a lifestyle built around the acquiring of an endless series of expensive objects is to reestablish the humbler but far richer notion of building our lives around other people.

"The need to belong," as the psychoanalyst and social philosopher Eric Fromm has pointed out, is central to all cultures and all times. Like Iceland we need to switch off the TV set at least one day a week. We need to exchange the secondhand life of watching other people live on the other side of a plastic screen with going out more and living our own "firsthand" lives—lives in which we meet with, work with, and play with other people. What better place to start this than in the community reuse and repair center?

We saw in chapter 10 how Gothenburg, Sweden, has set up a municipal reuse park that aims to be a fun place to visit for all the members of the family. A live band welcomes new visitors; clowns are on hand to amuse the children; curious onlookers can watch a dog that has been trained to separate mixed materials into six different categories; teenagers can attend rock concerts for an entrance fee of 20 plastic PET bottles; local TV personalities are on hand to answer questions from the public on everything from cooking to antiques; the toilets sport an exhibition of paintings—and in addition to all this are displays of donated reusable items from furniture to appliances to building materials that are laid out in exquisite fashion. This is truly a shopping mall for secondhand goods that everyone enjoys visiting.

We Need Our Artists and Communicators

The first five steps to zero waste involve developing new habits. But when we get to Steps 6–9 we are confronting the whole issue of consumerism; we are talking about changing a mind-set. Changing mind-sets is far more difficult than changing habits, and that is why we need our best artists, poets, writers, musicians, and other communicators to help us achieve zero waste. I was very happy in a recent visit to Marineo, a zero waste community in Sicily, to be presented with a CD of a song about zero waste by local composer Giuseppe De Sosa and to witness a delightful concert given by musicians aged six to fifteen from the musical academy in Palermo. We need to draw on all our creativity and communication skills to transform our society.

Make Love, Not Waste

Even in a world of dwindling finite material resources we have an infinite supply of love and affection, and we need to apply this even more diligently today than we have ever done before. In the 1960s the Vietnam peace movement came up with the slogan "Make Love, Not War." Now that we have to wage war against waste we might consider a modification, "Make Love, Not Waste"—or, less demanding, "Make Friends, Not Waste."

More Zeros on the Road to Sustainability

Efforts are underway in Italy to bring at least three movements together: zero waste, zero kilometers, and zero emissions. These are aimed at decentralized alternative energy production (zero emissions) and slow food, locally produced (zero kilometers). The goal would be to create communities where all three "zeros" are in play—a dramatic step toward sustainability and resilience.

Putting the World Together

I will end with a charming little story told to me many years ago by my Sunday school teacher. A clergyman was busy preparing his sermon, but his little grandson was tearing around the house making lots of noise. To quieten him down he took a map of the world and tore it up into tiny pieces and asked his grandson to reassemble it. To his amazement the grandson was back in rather a short time with the task complete. "How did you do that so quickly, Johnnie?" he asked.

"It was easy, Grandpa. On the back of the map was a picture of a man. As I put the man together the world fell into place." The clergyman had his sermon.

APPENDIX A

Sample Questions for Incinerator Hearings

The following are questions that Neil Carman, a former inspector of the Texas Air Control Board who often lends his expertise to communities fighting incinerators, prepared for citizens in Arecibo, Puerto Rico, to use at a public hearing addressing the environmental impacts of a proposed incinerator. These questions have been reproduced with his permission as a guide for citizens everywhere who find themselves fighting incinerator proposals but who can't afford to bring in expertise.

When commenting on incincrators, Carman argues that "the modeling of the state-of-the-art incinerator presents a false sense of security." He explains, "Modeling tends to only evaluate normal emissions and modeling routinely fails to consider worst-case upsets which could spike emissions hundreds or even thousands time higher than theoretical estimates based on ideal emission numbers."

1. What are the potential major upset conditions that could happen at the incinerator?
2. What would be the worst-case major upset event such as an explosion and fire that have occurred at many other state-of-the-art incinerators in the United States and worldwide?
3. What types of emissions would be produced during major upset events and malfunctions? Unburned gases of VOCs and dioxins? Higher particulate matter such as PM10, PM2.5, and ultrafines (nanoparticles)?
4. What potential volume of unburned and uncontrolled emissions could occur due to major upset events in the incinerator?
5. Does the incinerator possess a vent stack or bypass vent stack that will be used during emergencies when the air pollution control system needs to be bypassed?

    If the incinerators have bypass vent stacks, raise these concerns:

    A. The air modeling completely failed to review and consider bypass vent stack operations.
    B. No discussion was presented on the types of major upsets that result in use of the bypass vent stack.

C. No data or information was presented on the duration of bypass vent stack openings.

D. No data or information was presented on the volume of emissions from bypass vent stack openings.

E. No data or information was presented on the kinds of toxic and criteria pollutants released during bypass vent stack openings.

F. What emissions monitoring will be conducted during bypass vent stack openings?

G. What permit limits are placed on the bypass vent stack openings?

6. What is the permit limit on the number of major upset events each year in the incinerator? (Probably there is no limit, but some permits do specify limits.)

7. What is the permit limit on the length or duration of a major upset event in minutes and hours? (Permits usually lack such limits.)

8. What is the permit limit on the maximum allowable emissions rates during major upset events for VOCs, polycyclic aromatic hydrocarbons (PAHs), CO, PM10, PM2.5, ultrafines, metals, hydrogen chloride (HCl), hydrogen fluoride (HF), and other air toxics? (Usually the permit lacks any limits of emissions during upset events.)

9. Will there be a community acid gas related corrosion monitoring program to monitor corrosion caused by the release of acid gases such as HCl, HF, sulfuric acid (H_2SO_4), and other acid gases? Usually no such corrosion monitoring exists. But acid emissions from the incinerator will damage and corrode structures in the surrounding community. (Request that a community corrosion monitoring program be required in the permit.)

10. Will the permit require video monitoring of the smokestacks so that major upsets are videotaped and monitored? (The public should have access to video of the plant operations.)

11. Will the smokestack opacity limit be zero emissions of soot? (Any soot pollution level above 0 percent opacity will result in community impacts from the soot; during major upsets opacity may spike above the permit limit to 50–100 percent opacity.)

12. Will the community be allowed to track the opacity monitoring data?

13. How much of the incinerator's daily operational and monitoring data will be placed on real-time online access so that the community can track the emissions from the incinerator?

14. Will the smokestack have a dioxin continuous sampling and analysis system (e.g., the AMESA system) to determine on a near-real-time

basis the dioxin emissions since such systems have been in place in many European incinerators since the 1990s?

15. Air modeling is not a worst-case analysis of highest emissions such as fires mentioned under the emergency equipment section.

16. Air modeling did not consider or evaluate major upset events and is seriously flawed from a public health perspective since all incinerators experience major upset air pollution events when emissions of nearly all pollutants are often many times above the permit's maximum allowable emission rate limitations. The air modeling completely failed to review major upset conditions and no discussion was presented on the types of major upsets, the duration of major upsets, the kinds of toxic and criteria pollutants, or the volumes of emissions. The air modeling is a pie-in-the-sky scenario of smooth incinerator operations, when this is not real-world activity based on all other incinerators.

17. The permit lacks any stack continuous emissions monitoring system (CEMS) for the toxic acid gases of HCl, H_2SO_4, and HF. This is unacceptable. There will be no method of continuous compliance demonstration that the incinerators are meeting these limits.

18. The total tons of acid gases (HCl, H2SO4, and HF) that will be released by the proposed Arecibo incinerator are 39.8 tons a year, which equates to 79,600,000 pounds. The release of this amount of these highly acidic chemicals into the local environment is unacceptable because it will exacerbate respiratory problems among the local population.

19. Lead emissions from the incinerator pose a major health concern since we already have too much lead in area homes and the lead baseline used in the air modeling did not consider existing lead contamination problems in the community. The lead air modeling is completely flawed and inadequate for protecting the community.

20. We need a continuous lead ambient air-monitoring program to track the incinerator lead emissions in the community.

21. We need more blood lead monitoring and testing in our children.

22. Zinc oxide is produced by burning tires due to the high levels of zinc present in tires, and we are concerned about zinc oxide air emissions from the incinerator. No ambient air standards exist for zinc oxide, but this pollutant has been measured downwind of tire-burning sites such as industrial boilers. We request that a zinc oxide monitoring system be set up in the community.

23. The proposed incinerator project is far from state-of-the-art because we do not believe that the community's health will be protected by these facilities. (In the Arecibo application the allowed emissions are unacceptable and outrageously high even if they are met.)

24. Covanta Energy (and the incineration industry in general) has a poor environmental compliance record at its other state-of-the-art incinerators in the United States, and we request that a comprehensive compliance history review be conducted of Covanta Energy's poor history for the last fifteen years at every incinerator site. We understand that Covanta Energy has hundreds and potentially thousands of violations of its incinerator permit emissions limits and permit conditions. Why has no comprehensive review of Covanta Energy's compliance history been conducted so far? Where are the compliance records? Will they be produced for the community? Will Covanta Energy be transparent and show us their history at other incinerator sites?

25. We view this Covanta Energy incinerator with large volumes of toxic pollution in our community as another example of environmental injustice and think regulators needs to do a much better analysis of the Covanta Energy application since we believe there are many flaws and holes in it. We continue to have major concerns with environmental injustice being perpetrated here in our community, and EPA has not done nearly enough to alleviate our concerns of disproportionate impacts since the incinerator will be built here.

26. In Arecibo Covanta Energy's two municipal waste incinerators plan to emit at least 1,796.52 tons a year of criteria and toxic air pollution into our community's air supply. That is equal to 3,593,040 pounds of air pollution and 9,843.94 pounds every day of operation.

27. We are concerned that Covanta Energy's actual emissions will be far higher depending on the number of major upset events, duration of major upset events, and the types of emissions from these major upset events. The potential for major upset events has been heavily ignored in the application and best available control technology (BACT) review.

28. What volume of the many tons of volatile organic compounds (VOCs) are polycyclic aromatic hydrocarbons (PAHs), which are known human cancer-causing agents like benzo[a]pyrene (B[a]P)?

29. Why does the permit not address PAHs, even though all incinerators produce PAHs and distribute their exceptionally toxic characteristics?

30. Why did the air modeling completely ignore highly toxic PAHs?

31. Why are there no real limits in the permit on PAHs, including B(a)P and related PAHs?

32. What is the potential for PAHs to be absorbed onto soot/PM2.5 fine particles?

33. The EPA is currently proposing a stricter annual standard less than 15 micrograms per cubic meter for PM2.5. This raises the concern that current PM2.5 standards are not protective of human health.

34. When the incinerators have upset conditions, what is the opacity/ soot limit that the smokestacks will have to meet, and why is that not proposed in the permit?

35. Why are no PM2.5 CEMS being required on the smokestacks as a demonstration of continuous compliance?

36. PM2.5/soot emissions are extremely toxic as follows:

Soot, known as fine particulate matter, is regulated as PM2.5 (meaning particles smaller than 2.5 microns in aerodynamic diameter). Note that we are not talking about yard dust or desert sand that is more like PM50–100 and much larger than PM2.5 soot particles.

Note that every industrial factory, especially older ones, emit soot or PM2.5 particles—and typically the older plants emit much higher volumes of soot. PM2.5 particles are combustion by-products, meaning they come from burning fossil fuels and trash in incinerators and also coal, crude oil, petroleum coke, diesel, fuel oils, and natural gas. Note that coal, crude oil, petroleum coke, diesel, and fuel oils are worse in releasing more PM2.5 soot particles than burning gas. Natural gas produces very little soot or PM2.5.

Incinerators are among the most toxic sources of PM2.5/soot fine particles.

Soot is highly toxic and greatly hazardous for two reasons:

Fine particles are a hazard because they are so tiny and microscopic that they easily penetrate deeply into lung tissues, including the alveolar sacs, where oxygen is exchanged for carbon dioxide. Air pollution is more damaging when it reaches deep inside the sensitive lung tissues. PM2.5 fine particles are extremely hazardous due to their microscopic characteristics. Although billions of soot particles are visible when they lump together during release from a smokestack or diesel tailpipe so that a cloud of soot appears for an instant, once the soot particles separate they are no longer visible to the naked eye.

Soot is highly toxic (we know part of the story here although not all of it) because of toxic substances that compose it, including carcinogens.

A. One type of soot compound is PAHs. All smoke or soot contains PAHs. They are a type of compound made of multiple benzene rings like benzo[*a*]pyrene, a supercarcinogen more potent than benzene itself.

The California Air Resources Board in 1996 classified about 40 PAHs as human carcinogens, including benzo[*a*]pyrene, although there are hundreds of other PAHs.

PAHs are a sign of poor combustion. So if coal and fossil fuels do not burn completely, some PAHs are produced in the cooling stack or tailpipe gases, and they stick together as soot.

B. Metals may also be present in the soot depending on the fuel.

One theory of why soot pollution is harmful is that it changes the blood so it becomes slightly thicker and a person is more prone to a heart attack.

Soot particles are associated with bad-air days in urban areas. More heart attacks, more asthma attacks, more breathing problems, and so on are linked to days with higher soot levels even below the EPA PM2.5 National Ambient Air Quality Standards (NAAQS).

When higher soot is measured by urban monitors for one in every six days, epidemiologists have found higher rates of premature deaths, heart attacks, hospital admissions for breathing problems, and asthma attacks.

The PM2.5 NAAQS is a daily/twenty-four-hour standard and an annual standard. Both need to be tightened to protect public health.

37. State-of-the-art means zero emissions, and these proposed incinerators emit far too much toxic and criteria air pollution to be state-of-the-art incinerators. Any pollution above zero means that regulators and Covanta Energy (and other incinerator companies) are putting the community at risk. The Covanta Energy incinerators are far from state-of-the-art.

38. The proposed emissions rates for the pollutants from the Covanta Energy (and other incinerator companies) incinerators are meaningless and exist

only on paper, because no incinerator in the United States has ever been able to meet the emissions rates listed in the permits. Incinerators have so many problems that this application by Covanta Energy (and other incinerator companies) and the regulatory review completely ignore the real-world problems frequently experienced by large municipal waste incinerators.

39. Stack testing of the proposed incinerator is a community concern because it most likely will be flawed, misleading, and not representative of all conditions the incinerators operate under—such as major upsets.

A. Trial burns or stack testing protocols typically are highly idealized incinerator operating scenarios and not representative since they completely fail to address major upset events because such trial burns will only be conducted during routine, smooth, normal operations. If a major upset event begins to occur during a trial burn, the trial burn is always stopped until the major upset event and high emissions incident is over. Then a new trial burn will be initiated, throwing away all major upset event emissions data.

B. Stack testing during trial burns is flawed due to the EPA protocols used where surrogate chlorinated chemicals are used in place of dioxins and dibenzofurans. However, the surrogate chlorinated chemicals injected into the incinerator system are not representative of the formation of dioxins and dibenzofurans from other chlorinated carbon compounds (this also includes dioxins and furans formed de novo from chlorine, carbon, oxygen, and hydrogen).

C. Stack testing during trial burns is also flawed since a relatively small volume of stack gases and particulate matter is collected— being at most 1 percent or less throughout the three sampling runs across the stack diameter. The tiny total sampled volume is not necessarily representative of the nonsampled gases and particulate matter escaping during the stack testing.

D. Other potential errors occur due to oversampling or undersampling at each sample point along the stack diameters. If the stack gas/particulate matter (PM) flow rate is slightly higher and there is undersampling at a sample point, the sample reflects a lower pollutant catch compared to the overall stack pollutant flow rates.

E. Trial burns or stack testing protocols completely ignore vent bypass emissions events since no sampling is typically conducted during these operations.

F. Other errors and flaws may exist in the trial burns or stack testing protocols besides those mentioned here.

40. The permit does not appear to consider or address known incinerator emissions of ultrafines (nanoparticles) comprised of microscopic particulate matter smaller than one micron in diameter, since no federal regulatory standards yet exist for ultrafine particles. The community health concern about ultrafine particles is that they are even more toxic than regulated PM2.5/soot/fine particles and may represent a greater volume of toxic PM emissions than the PM2.5 fine particles. We request that the EPA regulate the ultrafine particles from the incinerator smokestacks and require Covanta Energy and other incinerator companies to consider installing air pollution capture devices to control for ultrafine particles.

41. Incinerators are well known to cause community health problems due to their significant volumes of toxic emissions and major upset emissions and malfunctions. The regulator needs to require a baseline community health study to document the health of local residents prior to their final permitting of the incinerator. A fatal flaw in incinerator regulations is the serious failure to require the conducting of a baseline community health study.

APPENDIX B

A Chronology of Zero Waste and Related Events

1974: The California Resource Recovery Association (CRRA) is founded. CRRA is one of the oldest recycling organizations in the United States. See www.crra.com

The Institute for Local Self-Reliance (ILSR) is founded in the Adams Morgan community of Washington, DC. ILSR's mission is to make this community of twenty-five thousand residents self-reliant. ILSR developed programs for waste utilization, urban food production, and decentralized energy systems. See www.ilsr.org.

1980: Urban Ore is begun by Daniel Knapp as a salvaging operation at the Berkeley Landfill. Eventually this becomes one of the world's largest and most successful reuse operations, with the stated mission "To End the Age of Waste." See a short history at http://urbanore.com/about-us/.

1986: The National Coalition Against Mass Burn Incineration and for Safe Alternatives is formed in Rutland, Vermont, by citizens fighting incinerator proposals in Connecticut, Massachusetts, New Hampshire, New York, and Vermont. Eventually, this cumbersome name morphs into Work on Waste USA, directed by Paul Connett. See www.americanhealthstudies.org.

1987: Sound Resource Management Group, a consulting company, is set up in Seattle by Jeffrey Morris and others to provide economic analysis on waste and other issues facing local governments. Since then it has been working to shrink pollution footprints, reduce waste, and conserve resources throughout the United States and Canada. The stated goal on its website today is "to help you achieve sustainability and zero waste in the 21st century." See www.zerowaste.com/pages/about.htm. In 1992, SRMG plays a prominent role in the design of the Nova Scotia waste program.

1988: Paul Connett delivers the Piskor lecture at St. Lawrence University in Canton, New York. This lecture, entitled "Waste Management as if the Future Mattered," is later circulated as a forty-eight-page booklet and a videotape.

In late 1988/early 1989 Seattle introduces curbside recycling and composting collections coupled with a pay-as-you-throw (PAYT) system for the residential fraction. Residual waste disposal drops dramatically.

1988–92: A majority of US states adopt recycling and waste diversion goals. During the same time Europe and Canada are developing extended producer responsibility (EPR) policies as ways to eliminate waste at the source. It will be more than another decade (2004) before EPR laws begin to be adopted in US states.

1989: The California legislature passes the California Integrated Waste Management Act (AB 939). This law requires all communities to reach a diversion of 25 percent from landfill by 1995 and 50 percent by 2000 and establishes a fine of ten thousand dollars a day for communities that do not reach the deadlines.

ILSR publishes their first of a series of important reports on the state of recycling in the United States by Brenda Platt and others, *Beyond 25 Percent: Materials Recovery Comes of Age.*

1991: ILSR publishes *Beyond 40 Percent: Record-Setting Recycling and Composting Programs.*

1992: ILSR publishes the third of their series of important reports on the state of recycling in the United States by Brenda Platt and others, *In-Depth Studies of Recycling and Composting Programs: Designs, Costs, Results.*

A study authored by Jeffrey Morris, *Review of Waste Management Options*, prepared for the city of Halifax in March, becomes the basis for Nova Scotia's successful waste program.

Alan Durning's book *How Much Is Enough?* is published by W. W. Norton as part of the Worldwatch Institute series of publications. A short essay summarizing this text can be accessed at www.thesocialcontract.com/pdf/three-three/Durning.pdf.

1993: To counter the inflated claims from the incinerator industry about the energy produced from their facilities, Jeffrey Morris published this important comparative analysis *Recycling versus Incineration: An Energy Conservation Analysis.* For more recent analysis contact Jeffrey Morris at jeff.morris@zerowaste.com.

Harold Crooks publishes *Giants of Garbage: The Rise of the Global Waste Industry and the Politics of Pollution Control* and coins the phrase "municipal industrial complex."

1994: The Western Australia Local Government Association adopts a policy statement on EPR that calls it "a pre-condition for a waste-free society."

1995: The US-based GrassRoots Recycling Network (GRRN) is cofounded by representatives of the Sierra Club, CRRA, and ILSR. Bill Sheehan is network coordinator and later executive director.

The Australian Capital Territory (ACT) government in Canberra is developing its "No Waste by 2010" bill.

Daniel Knapp brings news of the pending Canberra "No Waste by 2010" bill to the United States. Bill Sheehan circulates news. Lynn Landes sets up a website for Zero Waste USA; the emphasis is on changes at the individual level.

Paul Hawken publishes *The Ecology of Commerce: A Declaration of Sustainability*. Ray Anderson (CEO of Interface) reads this and sets out to make his company the first truly sustainable multinational corporation in the world.

Lois Gibbs publishes *Dying from Dioxin: A Citizen's Guide to Reclaiming Our Health and Rebuilding Democracy*.

1996: Canberra's "No Waste by 2010" bill is passed by the ACT government and becomes law.

First meeting of the GRRN in Pittsburgh, Pennsylvania, with sessions featuring zero waste.

Theo Colborn, Dianne Dumanoski, and John Peterson Myers publish *Our Stolen Future: Are We Threatening Our Fertility, Intelligence and Survival? A Scientific Detective Story*.

1997: Working with Bill Sheehan and the GRRN, Georgia Senator Donzela James introduces the first US legislation setting a goal of zero waste by 2020.

In April at the Rock Eagle 4-H Camp outside of Atlanta, Georgia, more than fifty people attend a meeting organized by the GRRN. Key outputs of the meeting are two resolutions to be adopted by organizations and local governments, one on zero waste (http://archive.grrn.org/zerowaste/zwmodel.html) and the other on producer responsibility (http://archive.grrn.org/resources/model_res_prod_resp.html). Those documents establish the boundaries of the US zero waste movement over the next decade and a half.

Newsweek features the US zero waste movement in the article "When You Can Really Make a Silk Purse from a Sow's Ear."

The CRRA holds its first conference devoted to zero waste in Monterey, California. The conference closing speaker is Jerry Brown (now the governor of California), who talks about the zen of zero waste. Over one thousand people attend, and the event is a major success. At this conference, the CRRA adopts the Agenda for the New Millennium drafted by Tedd Ward (the zero waste coordinator for Del Norte County) that calls for zero waste, ending welfare for wasting, and creating jobs from better product design and productive use of discards.

A new technical council is formed under the CRRA, the Global Recycling Council (GRC), to make sure the Agenda for the New Millennium is pursued by the CRRA after the conference.

Gary Liss, in his keynote address at the National Recycling Congress (the annual conference of the National Recycling Coalition) says "Every day, more than 100 million citizens do the right thing—they recycle. Now it is time to set our sights higher and start planning for the end to wasting resources and to our reliance on landfills, incinerators, and other waste facilities. Zero waste is a policy, a path, a direction, a target; it's a process, a way of thinking, a vision."

The TV documentary *Affluenza* is broadcast on PBS stations (producers John De Graaf and Vivia Boe). This is followed in 1998 with *Escape from Affluenza* (also produced by John De Graaf and Vivia Boe for PBS).

1998: The town of Carrboro, North Carolina, adopts both of GRRN's model resolutions for zero waste and EPR.

Seattle, Washington, includes zero waste as one of its guiding principles.

The National Recycling Coalition holds a conference in Albuquerque, New Mexico. Gary Liss organizes a session on zero waste businesses, highlighting the following examples: the Ontario container reuse program; Mad River Brewing Company; and Hewlett-Packard in Roseville, California.

Paul Connett writes *Municipal Waste Incineration: A Poor Solution for the Twenty-First Century* and presents the paper to the Fourth Annual International Management Conference in Amsterdam. This paper has since been translated into several languages and was the subject of two conferences in Japan orga-

nized by the incinerator industry (http://home.myfairpoint.net
/vzeeai8y/Poorsolution.pdf).

1999: The CRRA makes zero waste the theme of its annual conference,
held that year in June in San Francisco. Paul Connett videotapes
many of the participants at this CRRA conference. These include
Richard Anthony, Mary Lou Van Deventer, Daniel Knapp, Gary
Liss, Neil Seldman, Bill Sheehan, Tedd Ward, and John Young.
The video based on these interviews, *Zero Waste: Idealistic Dream
or Realistic Goal?*, is released in September 1999.

Robin Murray publishes his book *Creating Wealth from Waste*;
about one-third of the text is devoted to zero waste.

GRRN publishes *Welfare for Waste: How Federal Taxpayer
Subsidies Waste Resources and Discourage Recycling*, (www.grrn.org
/assets/pdfs/wasting/w4w.pdf).

Brenda Platt and Kelly Lease publish *Cutting the Waste Stream
in Half: Community Record-Setters Show How.*

The Philippines becomes the first country in the world to ban
incineration.

Zero waste is discussed in *Time*: Ivan Amato, "Can We Make
Garbage   Disappear?"   (www.time.com/time/magazine/article
/0,9171,992527,00.html).

2000: In February, Del Norte County, California, passes the United
States' first comprehensive zero waste plan.

Brenda Platt and Neil Seldman publish *Wasting and Recycling
in the United States 2000.*

In July, environmental activists from twelve Asia-Pacific
nations launch Waste Not Asia, the region's first alliance to
oppose the expansion of waste incineration technologies and
promote ecological methods of waste management. Waste
Not Asia clarifies that its alliance members will strive to put
in place a sustainable society that will constantly endeavor to
achieve a goal of zero waste through an evolving program of
clean production. The alliance's work will be based on principles
that emphasize materials recovery over materials destruction;
solutions that are democratically derived and socially just;
and systems that are community-based and emphasize local
jobs creation involving small businesses as opposed to capital-
intensive corporate-led interventions.

Nova Scotia achieves a 50 percent diversion from landfills and incinerators (in five years) using curbside collection of compostables and recyclables and other measures. Nova Scotia is the first province in Canada to reach this goal.

Paul Connett produces the video *On the Road to Zero Waste, Part 1: Nova Scotia.*

In October, an important meeting is held by GRRN in Blackhawk, Colorado, with many key players in zero waste (Annie Leonard, Peter Montague, Neil Seldman, Brenda Platt, Bill Sheehan, Eric Lombardi, Gary Liss, Helen Spiegelman, Rick Best, Warren Snow, and Paul Connett) involved.

Warren Snow proposes the idea of a Zero Waste International Alliance (ZWIA) at a breakout workshop. Twelve outcomes are proposed by the GRRN board over the next twelve months, one being to create the ZWIA. The goal is to establish an alliance of people working in the zero waste arena with a rotating secretariat— with GRRN taking on the role in year one. Snow is given the lead for this project to seek funding and coordinate establishment. At this meeting a ZWIA listserv is created with ten initial members (Richard Anthony, Linda Christopher, Michael Jenssen, Annie Leonard, Emily Miggins, Peter Montague, Bill Sheehan, Coy Smith, Warren Snow, and Tedd Ward).

In November, a ZWIA discussion document written by Warren Snow is released to the ZWIA listserv. Discussion on this and further steps needed to establish ZWIA continue on the listserv.

In December, a zero waste conference is held in Kaitaia, New Zealand. At this conference a ten-point action plan for the development of zero waste in 2001 is developed by Daniel Knapp (USA), Eric Lombardi (USA), Robin Murray (UK), Andy Moore (UK), and Mal Williams (UK). One of the action points is to get ZWIA off the ground (stalled due to lack of time and financial resources). Mal Williams (Wales) volunteers to take the lead on this.

In December, the Global Alliance for Incinerator Alternatives (GAIA) is founded at a meeting held in South Africa, with the participation of more than eighty people from twenty-three countries. Since this meeting, GAIA has grown to include more than 650 members in over ninety countries.

2001: The California Integrated Waste Management Board (CIWMB) adopts both zero waste and producer responsibility as goals in its 2001 strategic plan.

GRRN publishes a booklet by Bill Sheehan and Paul Connett entitled, *A Citizen's Agenda for Zero Waste: A United States/ Canadian Perspective.*

Bill Sheehan, executive director of GRRN, obtains a grant from the Giles W. and Elise G. Mead Foundation to research zero waste businesses and develop a certification process. Gary Liss does the research and is encouraged through that to develop zero waste business principles to guide businesses in a positive direction, without trying to develop a full certification program. GRRN adopts these zero waste business principles in January 2004.

At its May Blue Mountain retreat, GRRN declares, "Our mission is eliminating waste of natural resources, and in the process, waste of human resources. The most critical missing element needed to make progress towards a zero waste society at this point in time is extended producer responsibility for waste (EPR)."

In March, an interim ZWIA steering committee forms with twelve representatives from eight countries.

Zero waste is the subject of the cover story of the March/April issue of *E—The Environmental Magazine.*

In November, a letter is sent to ZWIA members from Zero Waste New Zealand proposing that the first ZWIA summit be held in New Zealand in April 2002. There is an enthusiastic response to this proposal, but fund-raising proves too difficult for many members to make this happen.

2002: Richard Anthony organizes (and Bill Sheehan secures funding for) the appearance of several zero waste activists at the R-02 conference in Geneva (Rick Anthony, Paul Connett, Daniel Knapp, Bill Sheehan, Bill Worrell, Joan Edwards, and Jeffrey Morris). Several of these experts go on to appear at a zero waste conference held at Sussex University and sponsored by CATS and Greenpeace UK. This conference is followed the next day with a meeting involving some of the guest speakers above, including Richard Anthony and Paul Connett, as well as Ralph Ryder from CATS, Robin Murray, and Mal Williams (Cylch recycling of Wales). Zero Waste United Kingdom (ZWUK) is set up. This meeting is retrospectively christened by Rick Anthony as the first zero waste dialogue.

Robin Murray publishes a second book dealing with zero waste, entitled simply *Zero Waste* and published by Greenpeace.

Michael Braungart and William McDonough publish the very influential book *Cradle to Cradle: Remaking the Way We Make Things*.

From November 2002 to January 2003, Warren Snow, using funding from UK sources, attempts to progress the ZWIA project. Julie Dickinson resigns as manager of Zero Waste New Zealand to help set up ZWIA. A New Zealand trust fund is set up to enable this on the understanding that it might only be temporary until an international structure for ZWIA is formed. Warren Snow and Julie Dickinson are legal trustees for the purpose of managing the day-to-day affairs of this trust fund.

2003: GAIA publishes *Waste Incineration: A Dying Technology*, written by Neil Tangri.

In March, a conference is held by GAIA in Penang, Malaysia. A ZWIA presentation is made to the conference and a special workshop held to discuss synergies and potential cooperation between ZWIA and GAIA. Richard Anthony has called this workshop the second zero waste dialogue.

In June, a ZWIA planning group is voted in by ZWIA list-serv to guide the future direction and purpose of ZWIA. Twelve members from seven countries are appointed.

On July 1, the Institute of Medicine publishes a report, *Dioxins and Dioxin-like Compounds in the Food Supply: Strategies to Decrease Exposure*.

The Product Policy Institute (PPI) is cofounded in July 2003 by Bill Sheehan and Helen Spiegelman to research and promote producer responsibility as an essential precondition for zero waste.

In October, funding is obtained by Mal Williams to host the first meeting of the ZWIA planning group and launch ZWIA in the United Kingdom. On October 19, the first official meeting of the ZWIA Zero Waste International planning group is held at the Bulkeley Hotel, Beaumaris, Wales. After this meeting, some of the attendees go on to speak at the Welsh Recycling Conference in Llandudno, Wales, and hold dialogues in Cardiff, Wales; Cork, Ireland; Doncaster, England; and Edinburgh, Scotland.

GRRN produces the video *On the Road to Zero Waste, Part 2: Burlington, Vermont* and *Part 3: Canberra, Australia* (www.american healthstudies.org).

2004: In January, GRRN adopts the zero waste business principles developed by Gary Liss.

The third zero waste dialogue takes place in San Francisco.

Paul Connett produces the video *On the Road to Zero Waste, Part 4: San Francisco* (www.americanhealthstudies.org/video.html#waste).

2005: In March, ZWIA adopts the slightly revised GRRN's zero waste business principles after much input over the ZWIA listserv.

2006: In June (in lieu of the flood of proposals to build facilities in North America claiming not to be incinerators) Greenaction and GAIA publish the report *Incinerators in Disguise: Case Studies of Gasification, Pyrolysis, and Plasma in Europe, Asia, and the United States* (www.greenaction.org/incinerators/documents /IncineratorsInDisguiseReportJune2006.pdf).

At the CRRA conference held in San Jose, Gunter Pauli speaks about his Zero Emissions Research Initiative (ZERI).

2007: Annie Leonard releases the video *The Story of Stuff*, which has now been translated into several languages and watched by millions of people around the world (www.storyofstuff.com/).

Zero waste is discussed in *Fortune* magazine: Marc Gunther, "The End of Garbage" (http://money.cnn.com/magazines/fortune /fortune_archive/2007/03/19/8402369/index.htm).

In April, Capannori (located near Lucca in Tuscany) becomes the first city in Italy to formally adopt a zero waste strategy.

In September, the fourth zero waste dialogue occurrs at the R-07 conference in Davos, Switzerland.

2008: The Sierra Club, the largest grassroots US environmental NGO, adopts a zero waste producer responsibility policy (www.sierraclub.org/policy/conservation/ZeroWasteExtended ProducerResponsibilityPolicy.pdf).

Capannori, Italy, hosts an international conference on zero waste with attendees from Belgium, Italy, Spain, the United Kingdom, and the United States. The attendees include Paul Connett, Richard Anthony, Eric Lombardi, and Jeffrey Morris from the United States.

In New Zealand the government passes the Waste Minimization Act. This law enshrines the zero waste philosophy as one of its goals.

The Irish government puts a fifteen-cent tax on plastic shopping bags. In one year the use of these bags falls by 92 percent.

The city of Toronto, Canada, declares a zero waste policy.

Newsweek discusses zero waste: Anne Underwood, "10 Fixes for the Planet" (www.thedailybeast.com/newsweek/2008/04/05/10 -fixes-for-the-planet.html).

2009: In February, the fifth zero waste dialogue is held in Naples, Italy.

In June, professor Vyvyan Howard, a specialist in infant and fetal pathology, presents an important review paper on nanoparticles, *Particulate Emissions and Health,* at a hearing on an incinerator proposal in Ireland (www.cawdrec.com/incineration/CVH.pdf).

In November, the sixth zero waste dialogue is held at the Asturias Hotel, Puerto Princesa, the Philippines.

2010: The first zero waste research center is started in Capannori, Italy. It involves an analysis of the residual fraction, the production of a "little museum of bad industrial design," and virtual input from experts throughout Italy, including Professor Andrea Segrè (University of Bologna); Enzo Favoino (Scuola Agraria del Parco di Monza); Roberto Cavallo (Asti, Piedmont); and Paul Connett (United States). It also involves local designers, students, artists, and activists.

In June, the Welsh Assembly issues the overarching waste strategy document for Wales, a report entitled *Towards Zero Waste: One Wales, One Planet.*

In October, the seventh zero waste dialogue is held in Florianópolis, Santa Catarina, Brazil.

2011: The Tellus Institute with Sound Resource Management Group publishes *More Jobs, Less Pollution: Growing the Recycling Economy in the U.S.* (www.container-recycling.org/assets/pdfs/jobs/More JobsLessPollution.pdf).

Fortune magazine discusses Zero Waste: David Ferry, "The Urban Quest for 'Zero' Waste" (http://online.wsj.com/article /SB10001424053111904583204576542233226922972.html).

The Hotel Association of Capri (Italy) adopts a zero waste strategy.

In October, the newly elected mayor and vice mayor of Naples adopt a zero waste strategy for the city.

On October 6–9, Capannori hosts another big international conference on zero waste, which is attended by many of the mayors that have declared zero waste in Italy. One of the guest speakers is Jack Macy from San Francisco—who gives a detailed presentation of the city's zero waste program—and also the vice mayor of Naples, Tomasso Sodano, who speaks about the city's bold plan to adopt zero waste.

Harrisburg, Pennsylvania, files for bankruptcy because of the huge debt incurred by its municipal waste incinerator.

2012: The full-length documentary on waste, *Trashed*—produced by Blenheim Films and featuring Jeremy Irons as host—receives its premiere showing at the Cannes Film Festival.

On May 11–12, a zero waste conference sponsored by Zero Waste Europe, local activists, and local governments is held in San Sebastián, Spain.

In June, a conference is held by the newly formed US Zero Waste Business Council (USZWBC) in Santa Rosa, California. This is attended by many business representatives, local officials, and activists from California and beyond.

By July, eighty-seven communities in Italy have formally adopted a zero waste strategy.

2013: By the beginning of 2013 over 100 communities in Italy have declared a zero waste strategy

An eighth zero waste dialogue takes place in March 2013 in Berkeley, California.

NOTES

Preface

1. C. Hedges, "Growth is the Problem." Chris Hedges's Columns, Truthdig blog, posted on September 10, 2012, accessed July 26, 2013, http://www.truthdig.com/report/item/growth_is_the_problem _20120910/.
2. A. Paton, "The Challenge of Fear," *Saturday Review*, September 9, 1967: 46.

Chapter 1. The Big Picture

1. Worldwatch Institute, "Global Municipal Solid Waste Continues to Grow," press release, July 24, 2012.
2. E. Elert, "Daily Infographic: If Everyone Lived Like an American, How Many Earths Would We Need?" *Popular Science*, October 10, 2012, accessed April 7, 2013, www.popsci.com/environment/article /2012-10/daily-infographic-if-everyone-lived-american-how-many -earths-would-we-need.
3. Global Footprint Network Advancing the Science of Sustainability. Website, accessed May 11, 2013, www.footprintnetwork.org/gfn_sub .php?content=global_footprint.
4. J. de Graaf and V. Boe (producers), *Affluenza*, (KCTS/Seattle and Oregon Public Broadcasting, aired September 15,1997); J. De Graaf and V. Boe (producers), *Escape from Affluenza*, 1998, accessed May 10, 2013, http://www.pbs.org/kcts/affluenza/.
5. J. de Graaf, D. Wann, and T. H. Naylor, *Affluenza: The All-Consuming Epidemic* (San Francisco: Berrett-Koehler, 2001).
6. A. Leonard, *The Story of Stuff*, video accessed May 10, 2013, www.storyofstuff.com/.
7. P. Hawken, *The Ecology of Commerce: A Declaration of Sustainability* (New York: HarperCollins, 1995).

Chapter 2. Ten Steps Toward a Zero Waste Community

1. Dan Knapp explains the twelve categories in Paul Connett (producer), *Zero Waste: Idealistic Dream or Realistic Goal?* (Canton, NY: Grass Roots and Global Video, 1999), accessed May 10, 2013, www.american healthstudies.org/video.html#waste.

2. P. Connett (producer), *On the Road to Zero Waste, Part 4: San Francisco*, (Canton, NY: Grass Roots and Global Video, 2004), accessed May 10, 2013 www.americanhealthstudies.org/video.html#waste.

3. SFenvironment, A Department of the City and County of San Francisco, *Zero Waste: Sending Nothing to Landfill Is a Foreseeable Future*, SFenvironment website, accessed April 7, 2013, www.sfenvironment .org/our_programs/overview.html?ssi=3.

4. P. Connett (producer), *On the Road to Zero Waste, Part 4: San Francisco*.

5. Garden Organic, "Master Composters Program," accessed April 7, 2013, www.gardenorganic.org.uk/composting/mastercomposter.php.

6. R. Bailey and P. Connett (producers), *Community Composting in Zurich* (Canton, NY: VideoActive Productions, 1991). For copies contact Paul Connett, 104 Walnut Street, Binghamton, NY 13905, or pconnett@gmail.com.

7. A. Segre, personal communication, March 20, 2010.

8. P. Connett (producer), *On the Road to Zero Waste, Part 1: Nova Scotia* (Canton, NY: Grass Roots and Global Video, 2000), accessed May 7, 2013, www.americanhealthstudies.org/video.html#waste.

9. P. Connett (producer), *Zero Waste: Idealistic Dream or Realistic Goal?*

10. P. Connett (producer), *On the Road to Zero Waste, Part 2: Burlington, Vermont* (Canton, NY: Grass Roots and Global Video, 2000), accessed May 7, 2013, www.americanhealthstudies.org/video.html#waste.

11. ReSOURCE, accessed May 7, 2013, www.resourcevt.org/history.

12. P. Connett (producer), *On the Road to Zero Waste, Part 2: Burlington, Vermont*.

13. Richard Anthony, personal communication, February 20, 2010.

14. "LWaRB Pledges £8m to Set Up London Reuse Network," July 12, 2010, Lets Recycle website accessed May 10, 2013, www.letsrecycle .com/news/latest-news/councils/lwarb-pledges-ps8m-to-set-up -london-reuse-network.

15. Global Alliance for Incinerator Alternatives, "Zero Waste Is Threatened in Canberra," 2009, accessed March 30, 2013, http:// org2.democracyinaction.org/o/1843/t/3022/p/dia/action/public /?action_KEY=265.

16. P. Connett (producer), *On the Road to Zero Waste, Part 3: Canberra, Australia* (Canton, NY: Grass Roots and Global Video, 2004), accessed March 30, 2013, www.americanhealthstudies.org/video.html#waste and www.youtube.com/watch?v=zIScs_Kwn-0.

17. Recovered Resource blog, "SF Hits 80% Diversion on the Road to Zero Waste," October 5, 2012, accessed March 30, 2013, http://blog.recology.com/2012/10/05/sf-hits-80-diversion-on-the-road -to-zero-waste/.

18. E. Rosenthal, "Motivated by a Tax, Irish Spurn Plastic Bags," *New York Times*, February 2, 2008, accessed April 7, 2013, www.nytimes.com /2008/02/02/world/europe/02bags.html.

19. J. Lowy, "Plastic Left Holding the Bag as Environmental Plague: Nations Around World Look at a Ban," *Seattle Post-Intelligencer*, July 21, 2004, accessed April 7, 2013, www.commondreams.org /headlines04/0721-04.htm.

20. C. Goodyear, "San Francisco First City to Ban Plastic Shopping Bags," *San Francisco Chronicle*, March 28, 2007.

21. Effecorta, website accessed May 7, 2013, www.Effecorta.it.

22. The Beer Store is featured in Paul Connett (producer), *Target Zero Canada* (Canton, NY: Grass Roots and Global Video, 2000). For copies contact Paul Connett, 104 Walnut Street, Binghamton, NY 13905, or pconnett@gmail.com

23. The New Parents Guide, "Diapers, Diapers and More Diapers," accessed April 7, 2013, www.thenewparentsguide.com/diapers.htm.

24. R. Knox, "Towns Tilting to Pay-per-Bag Trash Disposal," *Boston Globe*, November 8, 2007.

25. R. Hanley, "Pay-by-Bag Trash Disposal Really Pays, Town Learns," *New York Times*, November 24, 1988.

26. Roberto Cavallo, personal communication, March 12, 2008.

27. P. Connett (producer), *On the Road to Zero Waste, Part 1: Nova Scotia*.

28. For the citizens' role in the development of the Nova Scotia program contact David Wimberly, davidwimberly@eastlink.ca.

29. City of Halifax Review of Waste Management Options, Halifax (N.S.) Task Force (Halifax: Sound Resource Management Group (SRMG), 1992), accessed May 11, 2013, http://books.google.com/books?id =lQWBNwAACAAJ&dq=inauthor:%22Halifax+(N.S.).+Waste +Management+Task+Force%22&hl=en&sa=X&ei=2z6OUbCPG 8nN0AH__YDoBw&ved=0CCgQ6AEwAA.

30. J. Morris, E. Favoino, E. Lombardi et al., "What Is the Best Disposal Option for the 'Leftovers' on the Way to Zero Waste?" (Boulder, CO: Eco-Cycle, March, 2013), accessed May 11, 2013, http://ecocycle.org/files/pdfs /best_disposal_option_for_leftovers_on_the_way_to_Zero_Waste.pdf.

31. W. McDonough and M. Braungart, *Cradle to Cradle: Remaking the Way We Make Things* (New York: Northpoint, 2002).

32. Clean Production Action, Clean Production Strategies series, June 2009, accessed April 7, 2013, www.cleanproduction.org /Publications.php.

33. A. McPherson, B. Thorpe, and M. Rossi, "How Producer Responsibility for Product Take-Back Can Promote Eco-Design" (Somerville, MA: Clean Production Action, April 2008), accessed March 30, 2013, www.cleanproduction.org/library/cpa _ecodesign_Apr08.pdf.

34. European Commission, "Waste Electrical and Electronic Equipment," accessed March 30, 2013, http://ec.europa.eu/environment/waste /weee/legis_en.htm.

35. Electronic TakeBack Coalition, "State Legislation: States Are Passing E-Waste Legislation," Details on the State Laws and Programs, 2013, regularly updated, accessed March 30, 2013, www.electronicstakeback .com/promote-good-laws/state-legislation/.

36. Xerox, "2010 Report on Global Citizenship: Waste Prevention and Management," accessed March 30, 2013, www.xerox.com/corporate -citizenship-2010/sustainability/waste-prevention.html.

37. EPA, Federal Register, Vol. 53, No. 168, 33345, Tuesday August 30, 1988, accessed May 7, 2013, www.zerowasteamerica.org /LandfillsFedRegEPA.htm.

Chapter 3. A Brief History of the Anti-Incineration Movement

1. B. Commoner, M. McNamara, K. Shapiro, et al., *The Origins of Chlorinated Dioxins and Dibenzofurans Emitted by Incinerators That Burn Unseparated Municipal Solid Waste and an Assessment of Methods for Controlling Them* (Flushing, NY: Center for the Biology of Natural Systems, 1984).

2. The Center for the Biology of Natural Systems (CBNS) is a research organization with considerable experience in the analysis of environmental, energy, and resource problems and their economic implications, website at http://cbns.qc.edu/cbns/.

3. K. Olie, P.L. Vermeulen, and O. Hutzinger, "Chlorodibenzo-p-dioxins and Chlorodibenzofurans Are Trace Components of Fly Ash and Flue Gases of Some Municipal Incinerators in the Netherlands," *Chemosphere*, 6 (1977): 455–459.

4. F. Hasselriis, "Relationship Between Combustion Conditions and Emission of Trace Pollutants," paper presented to the New York State Air Pollution Association, May 2, 1984.

5. P. Connett, "Incineration and Dioxin," *Current, the Journal of North Country Action* (June,1985). North Country Action, P.O. Box 5055, Potsdam NY 13676.

6. P. Connett, "Dioxin: The Watergate of Molecules," a paper presented at Clarion College, Pennsylvania, April 2, 1993.

7. V. Ozvacic, "A Review of Stack Sampling Methodology for PCDDs and PCDFs," *Chemosphere*, 15 (1986): 1173–78.

8. Work on Waste (WOW) Anti-Incineration Records, St. Lawrence University Libraries, collection #164, accessed April 7, 2013, www.stlawu.edu/library/node/1934.

9. The Center for Health, Environment and Justice (CHEJ) was founded by Lois Gibbs (of Love Canal fame) under a different name, Citizens Clearinghouse for Hazardous Waste (CCHW). Their mission is "mentoring a movement, empowering people, preventing harm," with a website at http://chej.org/.

10. A. Hay and E. Silbergeld, "Assessing the Risk of Dioxin Exposure," *Nature*, 315 (1985): 102–3.

11. A. Hay and E. Silbergeld, "Dioxin Exposure at Monsanto," *Nature*, 320 (1986): 569.

12. P. Connett and T. Webster, "An Estimation of the Relative Human Exposure to 2,3,7,8-TCDD Emissions via Inhalation and Ingestion of Cow's Milk," *Chemosphere*, 16 (1987): 2079–84.

13. T. Webster and P. Connett, "Critical Factors in the Assessment of Food Chain Contamination by PCDD/PCDF from Incinerators," *Chemosphere*, 18 (1988): 1123–29.

14. T. Webster and P. Connett, "Cumulative Impact of Incineration on Agriculture: A Screening Procedure for Calculating Population Risk," *Chemosphere*, 19 (1989): 597–602.

15. T. Webster and P. Connett, "The Use of Bioconcentration Factors in Estimating the 2,3,7,8-TCDD Content of Cow's Milk," *Chemosphere*, 20 (1990): 779–86.

16. T. Webster and P. Connett, "Estimating Bioconcentration Factors and Half-Lives in Humans Using Physiologically Based Pharmacokinetic Modeling, Part 1: 2,3,7,8-TCDD," *Chemosphere*, 23 (1991): 1763–68.

17. T. Webster and P. Connett, "Dioxin Emission Inventories: The Importance of Large Sources," *Chemosphere*, 37 (1998): 2105–18.

18. P. Connett and T. Webster, "An Estimation of the Relative Human Exposure to 2,3,7,8-TCDD Emissions via Inhalation and Ingestion of Cow's Milk."

19. M. S. McLachlan, "Accumulation of PCDD/F in an Agricultural Food Chain," *Organohalogen Compounds*, 26 (1995): 105; see also M. S. McLachlan, "A Simple Model to Predict Accumulation of PCDD/Fs in an Agricultural Food Chain," *Chemosphere*, 34 (1997): 1263–76.

20. E. Connett and P. Connett, "The Netherlands: Milk and Meat Products Contaminated by Dioxin from Solid Waste Incinerator," *Waste Not* #61, June 29, 1989.

21. P. Connett and R. Bailey (producers), *Europeans Mobilizing Against Trash Incineration* (Canton, NY: VideoActive Productions, 1991). For copies contact Paul Connett, 104 Walnut Street, Binghamton, NY 13905, or pconnett@gmail.com.

22. ENDS Daily, "Dioxin Alert Shuts French Waste Incinerators," January 29, 1998, accessed April 7, 2013, www.cqs.com/frmsw.htm; see follow-up study from 2004, accessed April 7, 2013, www.epa.ie/downloads /pubs/other/dioxinresults/name,12038,en.html.

23. Environmental Protection Agency (Ireland), *Dioxins in the Irish Environment: An Assessment Based upon Levels in Cow's Milk* (Wexford, Ireland: EPA, 1996), reference cited in 2000 follow-up study, accessed May 10, 2013, http://www.epa.ie/pubs/reports/other/dioxinresults /EPA_dioxin_levels_in_the_irish_environment_2000.pdf.

24. A complete listing of the incinerators defeated in the United States between 1985 and 1995 can be found in *Waste Not* issues #251–74 (1994). *Waste Not* was edited by Ellen Connett and published from April 1988 to October 2000 by Work on Waste USA, 82 Judson Street, Canton, NY 13617. Back issues of *Waste Not* are available at www.americanhealthstudies.org.

25. VideoActive Productions was a nonprofit video company set up by Roger Bailey and Paul Connett, both professors at St. Lawrence University at the time. Between them they produced over forty video-tapes on various aspects of waste management (including a ten-part series on dioxin) between 1986 and 1995. A listing and description of all these videos can be found in the Resources section. All the videos have now been transferred to DVD and copies can be obtained by

contacting Paul Connett at 104 Walnut Street, Binghamton, NY 13905, or pconnett@gmail.com.

26. P. Connett and R. Bailey (producers), *Waste Management: As If the Future Mattered* (Canton, NY: VideoActive Productions, 1998).

27. P. Connett, "Waste Management: As if the Future Mattered," Piskor lecture at St. Lawrence University, Canton, NY, 1988.

28. R. Bailey and P. Connett (producers), *Dioxin* (ten-part series) (Canton, NY: VideoActive Productions, 1992); this was filmed at the First Citizens Conference on Dioxin.

29. The First Citizens Conference on Dioxin was held in Chapel Hill, North Carolina, in 1992. P. Connett and B. Elmore (editors), *The Proceedings of the First Citizens Conference on Dioxin* (Canton, NY: Work on Waste USA, 1992); available from Paul Connett at 104 Walnut Street, Binghamton, NY 13905, or pconnett@gmail.com.

30. F. H. Tschirley, "Dioxin," *Scientific American*, 254(2), (1986): 29–35.

31. Paul Connett, "Dioxin: Letter to the Editor," *Scientific American*, 254(4), (1986): 4.

32. Paul Connett, "Municipal Waste Incineration: A Poor Solution for the 21st Century," paper presented to the Fourth Annual International Management Conference in Amsterdam, November 24–25, 1998, accessed April 7, 2013, http://home.myfairpoint.net/vzeeai8y /Poorsolution.pdf.

33. Global Alliance for Incinerator Alternatives, Indigenous Environmental Network, Movement Generation, Grassroots Global Justice Alliance, et al., "Open Letter," accessed April 7, 2013, http:// grist.org/climate-energy/2010-10-23-open-letter-to-1-sky-from -the-grassroots/.

34. Back issues of Peter Montague's newsletters can be found in the archives at www.rachel.org/.

35. R. Gottlieb, *Forcing the Spring: The Transformation of the American Environmental Movement* (Washington, DC: Island Press, 2005), 234.

36. Beveridge and Diamond PC, "Supreme Court Decision Alters Solid Waste Flow Control Jurisprudence," May 2, 2007, accessed April 7, 2013, www.bdlaw.com/news-173.html.

37. Delaware Riverkeeper, "Incineration Threat to the Delaware River: One Down, More to Come?" newsletter, accessed July 30, 2012, www.delawareriverkeeper.org/resources/Factsheets/Delaware_River _One_Down_More_to_Come.pdf.

38. M. Fletcher, "Harrisburg, Pennsylvania's Capital, Files for Bankruptcy," *Washington Post*, October 12, 2011.

39. Beveridge and Diamond PC, "Supreme Court Decision Alters Solid Waste Flow Control Jurisprudence."

40. P. Connett and R. Bailey (producers), *Europeans Mobilizing Against Trash Incineration* (Canton, NY:VideoActive Productions, 1991). For copies contact Paul Connett, 104 Walnut Street, Binghamton, NY 13905, or pconnett@gmail.com.

41. L. Frühschütz, *Bürgeraktion Das Bessere Müllkonzept Bayern* [The Better Waste Concept Bavaria] (Ulm: Universitätsverlag, 1989).

42. E. Connett and P. Connett, "Over One Million People in Bavaria Vote to Put an Anti-Incinerator Referendum on the Ballot," *Waste Not* #122, October 25, 1990.

43. M. Berthoud, "Final Treatment of MSW and C&I Waste in Germany and Neighbouring Countries: How to Cope with Emerging Overcapacities?" ISWA website, 2011, accessed April 7, 2013, www.iswa.org/uploads/tx_iswaknowledgebase/Berthoud.pdf; also see discussion at UK Waste Incineration Network (UKWIN), "Sita Discussion of European Incineration Overcapacity," newsletter, August 28, 2012, accessed April 7, 2013, http://ukwin.org.uk/2012/08/28/2011-sita-discussion-of-european-incineration-overcapacity/.

44. "Parliament Backs Resource-Efficiency Roadmap," May 29, 2012, accessed April 7, 2013, www.euractiv.com/specialreport-recycling-society/european-parliament-backs-resour-news-512965.

45. United Nations Environmental Program (UNEP), Dioxin and Furan Inventories: National and Regional Emissions of PCDD/PCDF (Geneva, Switzerland: UNEP, 1999), accessed May 8, 2013, www.ipen.org/ipepweb1/library/DioxinInventory.pdf.

46. N. Tangri, "The First Ever National Incineration Ban," *GAIA*, June 23, 1999, accessed April 7, 2013, http://lists.essential.org/dioxin-l/msg00580.html.

47. E. Connett and P. Connett, "Waste Not Asia," *Waste Not* #465, August 14, 2000.

48. "Beijing to Lead Nation in Trash Burning," *China Daily*, March 6, 2012, accessed April 7, 2013, www.china.org.cn/environment/2012-03/06/content_24815415.htm.

49. E. Balkan, "The Dirty Truth About China's Incinerators," *The Guardian*, July 4, 2012, accessed April 7, 2013, www.guardian.co.uk/environment/2012/jul/04/dirty-truth-chinas-incinerators.

50. Yu Dawei, "Trash Incineration to Double by 2015," *CaixinOnline*, December 9, 2011, accessed April 7, 2013, http://english.caixin.com /2011-12-09/100336539.html.

51. N. J. Themelis, testimony on Bill 5118, "Reclassification of Trash-to-Energy Facilities as Class I Renewable Energy Sources," before the Environment Committee of Connecticut Legislature, March 2, 2012, accessed April 1, 2013, www.seas.columbia.edu/earth/wtert/sofos /Connecticut_Testimony_Themelis.pdf.

52. The list of sponsors for the Columbia University Earth Engineering Center is accessible online, accessed April 1, 2013, www.seas.columbia .edu/earth/sponsors.html.

53. The list of research associates for the Columbia University Earth Engineering Center is accessible online, accessed April 1, 2013, www.seas.columbia.edu/earth/research_associates.html.

54. The mission statement for the Waste-to-Energy Research and Technology Council is accessible online, accessed April 1, 2013, www.seas.columbia.edu/earth/wtert/mission.html.

55. The list of sponsors for the Waste-to-Energy Research and Technology Council is accessible online, accessed April 1, 2013, www.seas.columbia .edu/earth/wtert/sponsor.html.

56. N. J. Themelis, "Wondering What Dioxin/Furan Actually Is and How to Control Them? Prof. Nickolas Themelis Gives the Answer," Waste-to-Energy Research and Technology Council, accessed April 1, 2013, www.seas.columbia.edu/earth/wtert/sofos/dioxin_Furans_and_APC.pdf.

57. The mission statement of the Energy Recovery Council is accessible online, accessed April 1, 2013, www.energyrecoverycouncil.org/about.

58. "Waste-to-Energy Produces Clean Renewable Energy," accessed April 1, 2013, www.energyrecoverycouncil.org/waste-energy-produces-clean -renewable-a2984.

59. S. Greenfield, "Officials Insist Incinerator Will not Bankrupt Frederick," *Gazette.Net: Maryland Community News Online*, October 25, 2011, accessed April 1, 2013, www.gazette.net/article/20111025 /NEWS/710259873/1009/officials-insist-incinerator-will-not -bankrupt-frederick&template=gazette.

60. D. Andrews, "Reasons to Stop the Carroll/Frederick Incinerator," *Chesapeake* [Maryland Sierra Club newsletter], December 2011, accessed April 1, 2013, http://maryland.sierraclub.org/newsletter /archives/2011/12/a_014.asp.

61. J. Ostrowski, "Selection Committee Picks Low Bidder to Build West Palm Beach Incinerator Plant," *Palm Beach Post*, March 16, 2011, accessed April 1, 2013, www.palmbeachpost.com/money/selection -committee-picks-low-bidder-to-build-west-1326353.html.

62. C. Vyhnak, "Durham Incinerator Clears Key Hurdle," The Star.com [Toronto], May 29, 2008, accessed April 1, 2013, www.thestar.com /News/GTA/article/432754.

63. Durham Environmental Watch, "Incineration: Not Safe, not Sensible—the Facts and Issues in Brief," leaflet, updated May 24, 2011, accessed April 1, 2013, www.durhamenvironmentwatch.org /incineration.htm.

64. GrassRoots Recycling Network, "Toronto Declares Zero Waste!" *GRRN Newsletter*, January 29, 2008, accessed April 1, 2013, http://archive.grrn.org/zerowaste/articles/toronto_zerowaste.html.

65. "Toronto Area Waste-to-Energy Plant Location Selected," *Solid Waste and Recycling*, January 15, 2008, accessed April 1, 2013, www .solidwastemag.com/news/toronto-area-waste-to-energy-plant -location-selected/1000074186/?issue=01152008.

66. J. Barber, "Durham Plays 'Willing Host' to Incinerator," *Toronto Globe and Mail*, August 9, 2007, accessed April 1, 2013, www.theglobeand mail.com/news/national/durham-plays-willing-host-to-incinerator /article724716/.

67. P. Gombu, "Incinerator Approved but Opposition Burns," The Star.com [Toronto], January 25, 2008, accessed April 1, 2013, www .thestar.com/news/gta/2008/01/25/incinerator_approved_but _opposition_burns.html.

68. C. Vyhnak, "Angry Protesters Disrupt Durham Incinerator Groundbreaking," The Star.com [Toronto], August 17, 2011, accessed April 1, 2013, www.thestar.com/news/article/1040649--incinerator -party-goers-jeered-by-protesters.

69. J. Kirby, R. Warnica, V. Gustavo, et al., "99 Stupid Things the Government Spent Your Money On (Part II)," *Maclean's*, January 6, 2012, accessed April 1, 2013, http://www2.macleans.ca/2012/01/06/99 -stupid-things-the-government-spent-your-money-on-2/.

70. L. Gasser, "Durham-York Incinerator Receives Environmental Assessment Approval," Prevent Cancer (Ontario) website, accessed May 8, 2013, http://preventcancernow.ca/durham-york-incinerator -receives-environmental-assessment-approval.

71. Greenaction and Global Alliance for Incinerator Alternatives, *Incinerators in Disguise: Case Studies of Gasification, Pyrolysis, and Plasma in Europe, Asia, and the United States* (Berkeley, CA: Global Alliance for Incinerator Alternatives, 2006), accessed April 1, 2013, www .greenaction.org/incinerators/documents/IncineratorsInDisguise ReportJune2006.pdf.

72. D. Reevely, "Plasco Executive Earlier Played Key Role on Top City Files," *Ottawa Citizen*, November 25, 2011, accessed December 10, 2011, www.ottawacitizen.com/news/Councillors+have+final+Plasco +deal/5767330/story.html.

73. J. Willing, "Council Approves Plasco Deal," *Ottawa Sun*, December 14, 2011.

74. A. J. Hawkins, "Talking Trash: Burn It," *Crain's New York Business*, August 26, 2012, accessed April 1, 2013, www.crainsnewyork.com /article/20120826/REAL_ESTATE/308269974#ixzz24fRGDNDl.

75. Ananda Tan, personal communication, April 15, 2012.

76. Ananda Tan, personal communication, May 15, 2012.

Chapter 4. Incineration: The Biggest Obstacle to Zero Waste

1. AEA Technology, *Waste Management Options and Climate Change: Final Report 2002* (Luxembourg: European Communities, 2001), accessed May 8, 2013, http://ec.europa.eu/environment/waste/studies /pdf/climate_change.pdf.

2. "Calvert Incinerator Plan Approved," *Bucks Herald*, July 29, 2012, accessed July 29, 2012, www.bucksherald.co.uk/news/local-news /calvert-incinerator-plan-approved-1-3759045.

3. N. Mann, "SITA UK Signs £1 Billion Suffolk Waste PFI Deal," Lets Recycle website, October 6, 2010, accessed July 29, 2012, www .letsrecycle.com/news/latest-news/councils/sita-uk-signs-ps1-billion -suffolk-waste-pfi-deal.

4. "Sita Jobs: Great Blakenham Waste Incinerator, 43 Job Vacancies," Job Vacancies website, accessed July 29, 2012, www.jobvacancies.org /sita-jobs-great-blakenham-waste-incinerator-43-job-vacancies.

5. "The Energy Way: Brescia's Waste Incinerator," *Centro Volta*, no date given, accessed July 29, 2012, www.centrovolta.it/laviadellenergia /inglese/produzione/brescia.htm.

6. Beppe Grillo blog, "See Brescia and Die," January 12, 2008, accessed July 29, 2012, www.beppegrillo.it/en/2008/01/see_brescia_and_die.html.

7. P. Connett (producer), *On the Road to Zero Waste, Part 1: Nova Scotia* (Canton, NY: Grass Roots and Global Video, 2000), accessed May 8, 2013, www.americanhealthstudies.org/video.html#waste.

8. R. Murray, *Creating Wealth from Waste* (London: Demos, 1999).

9. J. Morris and D. Canzoneri, *Recycling versus Incineration: An Energy Conservation Analysis* (Seattle, WA: Sound Resource Management Group, 1993). For more recent analysis contact Jeffrey Morris at jeff.morris@zerowaste.com.

10. Franklin Associates, *The Role of Recycling in Integrated Solid Waste Management to the Year 2000* (Stamford, CT: Keep America Beautiful, 1994).

11. ICF Consulting, "Incineration of Municipal Solid Waste: A Reasonable Energy Option?" Fact Sheet #3 (2005), accessed April 1, 2013, www.durhamenvironmentwatch.org/Incinerator%20Files /FS3energy.pdf.

12. "Panorama, Rubbish," BBC-1 TV, June 26, 2004, accessed April 1, 2013, http://news.bbc.co.uk/hi/english/static/audio_video /programmes/panorama/transcripts/transcript_26_06_00.txt.

13. S. Hanna, "Kent's Waste Contract Could Be Money in the Bin," Kent Online, August 12, 2008, accessed April 1, 2013, www.kentonline .co.uk/kentonline/newsarchive.aspx?articleid=46264.

14. C. Brady (director), *Trashed*, an environmental documentary feature film (Watlington, Oxfordshire, UK: Blenheim Films, 2012), accessed April 7, 2013, www.trashedfilm.com/about-trashed.php.

15. *Waste-to-Energy Report*, September 10, 1986. *The Waste-to-Energy Report*, a biweekly newsletter on resource recovery, was published by McGraw Hill until it was discontinued April 26, 1993; trademark details accessed May 10, 2013, https://www.trademarkia.com /wastetoenergy-report-73557452.html.

16. E. Kolbert, "Incinerators Held to Pose Ash Hazard," *New York Times*, September 30, 1987, accessed July 29, 2012, www.nytimes.com/1987 /09/30/nyregion/incinerators-held-to-pose-ash-hazard.html ?pagewanted=all&src=pm.

17. "Linguistic Detoxification," *Webster's Online Dictionary*, accessed July 29, 2012, www.websters-online-dictionary.org/definitions/Linguistic +detoxification?cx=partner-pub-0939450753529744%3Av0qd01-tdlq &cof=FORID%3A9&ie=UTF-8&q=Linguistic+detoxification&sa =Search#906.

18. E. Connett and P. Connett, "The Great Incinerator Ash Scam," *Waste Not* #316–19, March 1995.

19. L. Greenhouse, "Justices Decide Incinerator Ash Is Toxic Waste," *New York Times*, May 3, 1994, accessed July 29, 2012, www.nytimes.com /1994/05/03/us/supreme-court-roundup-justices-decide-incinerator -ash-is-toxic-waste.html?pagewanted=all&src=pm.

20. E. Connett and P. Connett, "The Great Incinerator Ash Scam."

21. L. Gibbs, *Dying from Dioxin: A Citizen's Guide to Reclaiming Our Health and Rebuilding Democracy* (Boston: South End Press, 1995).

22. E. Connett and P. Connett, "Mercury in Massachusetts: An Evaluation of Sources, Emissions, Impacts and Controls," *Waste Not* #363, Summer 1996.

23. K. Olie, P. L. Vermeulen, and O. Hutzinger, "Chlorodibenzo-p-dioxins and Chlorodibenzofurans Are Trace Components of Fly Ash and Flue Gases of Some Municipal Incinerators in the Netherlands," *Chemosphere*, 6 (1977): 455–459.

24. C. Brady (director), *Trashed*.

25. Neil Carman can be reached at neil_carman@greenbuilder.com.

26. E. Connett and P. Connett, "The Smoking Gun (Columbus, Ohio)," *Waste Not* #302, September 1994.

27. R. De Fre and M. Wevers, "Underestimation in Dioxin Emission Inventories," *Organohalogen Compounds*, 36 (1998): 17–20.

28. "Dioxin Monitoring: Amesa System," Amesa website, accessed July 29, 2012, www.environnement-sa.com/dioxin_monitoring-produits _amesa.html.

29. Environmental Protection Agency, *Reanalysis of Key Issues Related to Dioxin Toxicity and Response to NAS Comments:* Volume 1, February 17, 2012, accessed July 29, 2012, http://cfpub.epa.gov/ncea/CFM /nceaQFind.cfm?keyword=Dioxin.

30. "EPA Sets Safe Dioxin Level," *Chemistry World*, February 23, 2012, accessed July 29, 2012, http://cfpub.epa.gov/ncea/CFM/nceaQFind .cfm?keyword=Dioxin.

31. "EPA Dioxin Reassessment Report," Agent Orange Zone (blog), February 17, 2012, accessed July 29, 2012, http://agentorangezone .blogspot.com/2012/02/epa-dioxin-reassessment-report.html.

32. H. J. Pluim, J. G. Koppe, K. Olie, et al., "Effects of Dioxins on Thyroid Function in Newborn Babies," *The Lancet*, 339 (1992): 1303.

33. L. S. Birnbaum, "Developmental Effects of Dioxins."

34. T. Colborn, J. P. Myers, and D. Dumanoski, *Our Stolen Future: How Man-Made Chemicals Are Threatening Our Fertility, Intelligence and Survival* (New York: Dutton, 1996).

35. Institute of Medicine, *Dioxin-like Compounds in the Food Supply: Strategies to Decrease Exposure* (Washington, DC: National Academies Press, 2003).

36. N. J. Themelis, "Wondering What Dioxin/Furan Actually Is and How to Control Them? Prof. Nickolas Themelis Gives the Answer," Waste-to-Energy Research and Technology Council, accessed April 1, 2013, www.seas.columbia.edu/earth/wtert/sofos/dioxin_Furans_and_APC.pdf.

37. M. S. McLachlan, "Accumulation of PCDD/F in an Agricultural Food Chain," *Organohalogen Compounds*, 26 (1995): 105; see also M. S. McLachlan, "A Simple Model to Predict Accumulation of PCDD/Fs in an Agricultural Food Chain," *Chemosphere*, 34 (1997): 1263–76.

38. L. S. Birnbaum, "Developmental Effects of Dioxins," *Environmental Health Perspectives*, 103 (1985): 89–94, accessed July 29, 2012, http://ehp03.niehs.nih.gov/article/fetchArticle.action;jsessionid =1325FCDFBDC68562FB64BDE503B97775?articleURI=info %3Adoi%2F10.1289%2Fehp.95103s789.

39. J. Y. Wang, F. Liu, F. Jiao, et al., "Time-Dependent Translocation and Potential Impairment of Central Nervous System by Intranasally Instilled TiO2 Nanoparticles," *Toxicology*, 254 (2008): 82–90, accessed July 29, 2012, www.sciencedirect.com/science/article/pii/S0300483X08004411.

40. S. A. Cormier, S. Lomnicki, W. Backes, et al., "Origin and Health Impacts of Emissions of Toxic By-Products and Fine Particles from Combustion and Thermal Treatment of Hazardous Wastes and Materials," *Environmental Health Perspectives*, 114 (2006): 810–17, accessed July 29, 2012, http://ehp03.niehs.nih.gov/article/fetchArticle .action?articleURI=info%3Adoi%2F10.1289%2Fehp.8629.

41. R. Maynard and C. V. Howard (editors), *Particulate Matter: Properties and Effects upon Health* (Oxford: BIOS Scientific Publishers, 1999).

42. C.V. Howard, "Particulate Emissions and Health, Statement of Evidence Regarding Proposed Ringaskiddy Waste-to-Energy Facility," June 2009, accessed May 10, 2013, http://www.durhamenvironment watch.org/Incinerator%20Health/CVHRingaskiddy EvidenceFinal1.pdf.

43. C. A. Pope III, R.T. Burnett, G.D. Thurston, et al., "Cardiovascular Mortality and Long-Term Exposure to Particulate Air Pollution:

Epidemiological Evidence of General Pathophysiological Pathways of Disease," *Circulation*, 109 (2004): 71–77.

44. D. W. Dockery and P.H. Stone, "Cardiovascular Risks from Fine Particulate Air Pollution," *New England Journal of Medicine*, 356 (2007): 511–13.

45. Resource Recycling Solutions (Derbyshire) Ltd (RRS), "Development of Waste Treatment Facility, comprising Reception and Recycling Hall; Mechanical Biological Treatment (MBT) Facility; Advanced Conversion Technology (ACT) Facility; Power Generation and Export Facility; Education and Office Accommodation; Landscaping and Access. Sinfin Lane, Derby," prepared by Angela Baje and Patrick Froggatt (Leeds: RRS, May 2009), see Chapter 7 (Air Quality), paragraph 7.5.32., accessed May 11, 2013, http://eplanning.derby.gov.uk/acolnet/DocumentsOnline /documents/28590_49.pdf.

46. J. Tickner, C. Raffensperger, and N. Myers, "The Precautionary Principle in Action: A Handbook," Science and Environmental Health Network, accessed July 29, 2012, www.mindfully.org/Precaution /Precaution-In-Action-Handbook.pdf.

47. A complete listing of the incinerators defeated in the United States between 1985 and 1995 can be found in *Waste Not* issues #251–74 (1994). Back issues of *Waste Not* are available at www.american healthstudies.org.

48. T. H. Christensen (editor), *Solid Waste Technology and Management*, 2 vols. (New York: Wiley, 2010), 366; accessed April 12, 2012, http:// books.google.com/books?id=4gxbMOhpxEC&pg=PT320&lpg=PT320 &dq=Incineration+hamburg+1895&source=bl&ots=dAJR9uiZvC &sig=DE5r7cOrJRLg5xhR8Ki-ogD73C4&hl=en#v=onepage &q=Incineration%20hamburg%201895&f=false.

Chapter 5. The Early History of Zero Waste

1. A. Koestler, *The Ghost in the Machine* (New York: Penguin Arkana. 1967).
2. V. Packard, *The Waste Makers* (New York: D. McKay, 1960).
3. M. Ostrow (director), *Race to Save the Planet* (Annenberg PBS, 1990).
4. Charles Moore is founder of the Algalita Marine Research Foundation. He captains the foundation's research vessel, the *Alguita*, documenting the great expanses of plastic waste that now litter our oceans. He is featured in the documentary movie *Trashed*.

5. C. Brady (director), *Trashed*, an environmental documentary feature film (Watlington, Oxfordshire, UK: Blenheim Films, 2012), accessed April 7, 2013, www.trashedfilm.com/about-trashed.php.
6. R. Carson, *Silent Spring* (New York: Houghton Mifflin, 1962).
7. T. Colborn, J. P. Myers, and D. Dumanoski, *Our Stolen Future: How Man-Made Chemicals Are Threatening Our Fertility, Intelligence and Survival* (New York: Dutton, 1996).

Chapter 6. The United States

1. A complete listing of the incinerators defeated in the United States between 1985 and 1995 can be found in *Waste Not* issues #251–74 (1994). Back issues of *Waste Not* are available at www.american healthstudies.org.
2. J. Carroll, "U.S. EPA Changes National Recycling Rates," *Waste Recycling News,* February 13, 2012, accessed April 7, 2013, www .wasterecyclingnews.com/article/20120213/NEWS02/302139995 /u-s-epa-changes-national-recycling-rates.
3. R. W. Beck, *U.S. Recycling Economic Information Study* (Washington, DC: National Recycling Coalition, 2001), accessed April 7, 2013, www.epa.gov/wastes/conserve/tools/rmd/rei-rw/pdf/n_report.pdf.
4. "Integrated Waste Management Act, California (AB 939)," Californians Against Waste, accessed April 2, 2013, www.cawrecycles .org/facts_and_stats/california_recycling_laws.
5. "Integrated Waste Management Act, California (AB 939)," Californians Against Waste.
6. City and County of San Francisco, "Recycling Program: Mayor Announces City Leads America in Recycling," press release, October 9, 2012, accessed April 2, 2013, http://www5.sfgov.org/sf_news/recycling/.
7. R. Reed, Recology. Personal communication, May 10, 2013.
8. V. Gokaldas, "San Francisco, USA: Creating a Culture of Zero Waste," in C. Allen, V. Gokaldas, A. Larracas et al., *On the Road to Zero Waste: Successes and Lessons from Around the World* (Berkeley, CA: Global Alliance for Incinerator Alternatives, 2012), accessed April 2, 2013, www.no-burn.org/on-the-road-to-zero-waste-successes-and-lessons -from-around-the-world.
9. R. Reed, Recology. Personal communication, May 10, 2013.
10. V. Gokaldas, "San Francisco, USA: Creating a Culture of Zero Waste."
11. V. Gokaldas, "San Francisco, USA: Creating a Culture of Zero Waste."

12. V. Gokaldas, "San Francisco, USA: Creating a Culture of Zero Waste."

13. V. Gokaldas, "San Francisco, USA: Creating a Culture of Zero Waste."

14. V. Gokaldas, "San Francisco, USA: Creating a Culture of Zero Waste."

15. V. Gokaldas, "San Francisco, USA: Creating a Culture of Zero Waste."

16. G. Newsom and J. Blumenfeld, "Letter to Mayor Pietro Vignali of Parma," (undated) 2010, accessed May 9, 2013, www.gestione correttarifiuti.it/sito/uploads/Vignali.pdf.

17. V. Ibanez (director), *Zero Waste* (Italy: Ibanez, 2011), accessed May 9, 2013, www.cinemambiente.it/film_ambiente/5598_.html.

18. "California's New Goal: 75% Recycling," CalRecycle, May 9, 2012, accessed April 2, 2013, www.calrecycle.ca.gov/75percent/Plan.pdf.

19. P. Connett (producer), *Zero Waste: Idealistic Dream or Realistic Goal?* (Canton, NY: Grass Roots and Global Video, 1999), accessed May 9, 2013, www.americanhealthstudies.org/video.html#waste.

20. "Albertsons Earns Award for Zero Waste Stores," *Santa Barbara Independent*, December 15, 2011, accessed April 2, 2013, www .independent.com/news/2011/dec/14/albertsons-earns-2011 -governors-environmentalecono/.

21. "Albertsons Earns Award for Zero Waste Stores."

22. "Albertsons Earns Award for Zero Waste Stores."

23. "Albertsons Earns Award for Zero Waste Stores."

24. "Building Zero Waste Communities," Eco-Cycle, accessed April 2, 2013, www.ecocycle.org/. Eric Lombardi can be contacted at eric@ecocycle.org.

25. "Building Zero Waste Communities."

26. "Building Zero Waste Communities."

27. V. Garza, "Austin Approves 30-Year Plan to Reduce Waste by 90%," *Austin Business Journal*, December 16, 2012, accessed April 2, 2013, www.bizjournals.com/austin/news/2011/12/16/austin-approves-30 -year-plan-to-reduce.html.

Chapter 7. Italy

1. J. M. Simon, "Castelbuono: The Town Where Even Donkeys Walk the Zero Waste Path," Zero Waste Europe, April 4, 2012, accessed April 2, 2013, www.zerowasteeurope.eu/2012/04/castelbuono-the-town-where -even-donkeys-walk-the-zero-waste-path/.

2. J. M. Simon, "Castelbuono: The Town Where Even Donkeys Walk the Zero Waste Path."

3. J. M. Simon, "Castelbuono: The Town Where Even Donkeys Walk the Zero Waste Path."

4. P. Connett, R. Ercolini, and P. Lo Sciuto, *Rifiuti Zero: Una Rivoluzione in Corso* (Viareggio, Italy: Disessensi, 2012).

Chapter 8. Australia and New Zealand

1. The ACT no waste law was passed in 1996. The website for the administration of this law is www.tams.act.gov.au/live/recycling-waste /about_ACT_NOWaste.

2. P. Connett (producer), *On the Road to Zero Waste, Part 3: Canberra, Australia* (Canton, NY: Grass Roots and Global Video, 2004), accessed March 30, 2013, www.americanhealthstudies.org/video.html#waste and www.youtube.com/watch?v=zIScs_Kwn-0.

3. Global Alliance for Incinerator Alternatives, "Zero Waste Is Threatened in Canberra," newsletter, 2009, accessed April 7, 2013, http://org2.democracyinaction.org/o/1843/t/3022/p/dia/action/public /?action_KEY=265.

4. City News, "Canberra's Waste Dilemma" (Canberra, Australia: CityNews.com.au, March 2, 2011), accessed April 2, 2013, http:// citynews.com.au/2011/canberras-waste-dilemma/.

5. City News, "Canberra's Waste Dilemma" (Canberra, Australia: CityNews.com.au, March 2, 2011), accessed April 2, 2013, http:// citynews.com.au/2011/canberras-waste-dilemma/.

6. C. Allen, GAIA, personal communication, September 29, 2011. This information was contained in an early draft of the Global Alliance for Incinerator Alternatives report *On the Road to Zero Waste: Successes and Lessons from Around the World* (Berkeley, CA: Global Alliance for Incinerator Alternatives, 2012) but not used in the final version. The final report is available online at www.no-burn.org/on-the-road-to -zero-waste-successes-and-lessons-from- around-the-world.

7. Ministry of the Environment (New Zealand), *The Waste Minimization Act 2008*, accessed May 10, 2013, www.mfe.govt.nz/issues/waste/waste -minimisation.html; see also www.legislation.govt.nz/act/public /2008/0089/latest/viewpdf.aspx?search=ts_act_waste_resel.

8. C. Allen, GAIA, personal communication, September 29, 2011.

9. C. Allen, GAIA, personal communication, September 29, 2011.

10. W. Snow, personal communication, August 19, 2012.

Chapter 9. Canada

1. P. Connett (producer), *Target Zero Canada* (Canton, NY: Grass Roots and Global Video, 2000). For copies contact Paul Connett, 104 Walnut Street, Binghamton, NY 13905, or pconnett@gmail.com.

2. Environmental Protection Agency, *Federal Register*, 53(168) (August 30, 1988): 3334. Cited in S. C. Christenson and I. M. Cozzarelli, "The Norman Landfill Environmental Research Site: What Happened to the Waste in Landfills?" fact sheet, 2003, accessed April 7, 2013, http://pubs.usgs.gov/fs/fs-040-03/.

3. P. Connett (producer), *On the Road to Zero Waste, Part 1: Nova Scotia* (Canton, NY: Grass Roots and Global Video, 2000), accessed May 10, 2013, www.americanhealthstudies.org/video.html#waste.

4. J. Morris, *Review of Waste Management Options* [report for the City of Halifax] (Seattle, WA: Sound Resource Management Group, 1992).

5. J. Corson, *The Resourceful Renovator* (Toronto: Key Porter Books, 2000).

6. "Measuring Sustainable Development: Application of the Genuine Progress Index to Nova Scotia," Nova Scotia GPI Solid Waste-Resource Accounts, 2004, accessed May 10, 2013, www.gpiatlantic.org/pdf/solidwaste/solidwaste.pdf.

7. "Our Path Forward: Building on the Success of Nova Scotia's Solid Waste-Resource Management Strategy," Nova Scotia website, 2011, accessed May 10, 2013, www.gov.ns.ca/nse/waste/docs/Solid.Waste.Strategy-Our.Path.Forward.2011.pdf.

8. "Greening Our Waste for Guelph's New Organic Waste Processing Facility," City of Guelph website, accessed May 10, 2013, http://guelph.ca/?s=organic+waste+facility.

9. City of Guelph Solid Waste Management Master Plan—Waste Prevention and Diversion Best Practices Discussion Paper, Appendix C, Gartner Lee, 2008, City of Guelph: website, accessed May 10, 2013, http://guelph.ca/wp-content/uploads/SWMMP_AppendixC.pdf.

10. R. Bailey and P. Connett (producers), *WasteWise: A Community Resource Center* (Canton, NY: VideoActive Productions, 1991). For copies contact Paul Connett, 104 Walnut Street, Binghamton, NY 13905, or pconnett@gmail.com.

11. P. Hawken, *The Ecology of Commerce: A Declaration of Sustainability* (New York: HarperCollins, 1995); see also *The Ecology of Commerce: A Declaration of Sustainability*, rev. ed. (New York: HarperCollins, 2010).

12. R. Anderson, *Ray Anderson—Interface Carpet,* video statement, uploaded onto YouTube, October 25, 2007, accessed April 6, 2013, www.youtube .com/watch?v=OUG4JXE6K4A&noredirect=1; see also R. Anderson, "Green Business," TED talk, accessed April 6, 2013, www.grinningplanet .com/embed/green-business-video/interface-carpet-ray-anderson-ted.htm.

13. E. Connett, "Municipal Waste Incineration Banned in Ontario, Canada," *Waste Not* #146, April 10, 1991.

14. P. Connett (producer), *Target Zero Canada.*

15. E. Connett, "The Newly Elected Government in Ontario, Canada, Prepares to Lift the Ban on Building MSW Incinerators, Part 1," *Waste Not* #342, August 14, 1995.

16. E. Connett, "The Newly Elected Government in Ontario, Canada, Prepares to Lift the Ban on Building MSW Incinerators, Part 2," *Waste Not* #343, August 14, 1995.

17. Details of the Trenton program can be obtained from Quinte Waste Solutions, 270 West Street, Trenton, Ontario K8V 2N3 ([613] 394-6266), and online at www.toolsofchange.com/en/case -studies/detail/19.

18. GrassRoots Recycling Network, "Toronto Declares Zero Waste!" newsletter, January 29, 2008, accessed May 10, 2013, http://archive .grrn.org/zerowaste/articles/toronto_zerowaste.html.

19. "Toronto Area Waste-to-Energy Plant Location Selected," *Solid Waste and Recycling,* January 15, 2008, accessed April 7, 2013, www.solidwastemag.com/news/toronto-area-waste-to-energy -plant-location-selected/1000074186/?issue=01152008.

20. "Take It Back!" City of Ottawa website, accessed April 7, 2013, http://app01.ottawa.ca/takeitback/Welcome.do?lang=en; see also "About the Take It Back! Program" at http://app01.ottawa.ca /takeitback/Background.do?lang=en.

21. "A Description of Edmonton's Waste Program," City of Edmonton website, accessed April 7, 2013, www.edmonton.ca/environmental /documents/City_Environmental_Programs.pdf.

22. Enerkem, "Waste Becomes Fuels, Edmonton, Alberta," Enerkem website, accessed May 10, 2013, http://www.enerkem.com/en/facilities /plants/edmonton-alberta-canada.

23. B. West, "Incinerator Critic Not Alone," letter, *North Shore News* (Vancouver), August 26, 2011, accessed April 7, 2013, www.nsnews .com/technology/Incinerator+critic+alone/5310522/story.html

#ixzz1WB500En2. For more information contact Ben West, Wilderness Committee, Healthy Communities Campaigner at (604) 710-5340, or ben@wildernesscommittee.org.

24. H. Spiegelman, "Time for Metro Vancouver to Forget About Incineration, Get Going on Composting," Vancouver Food Policy Council blog, August 8, 2011, accessed April 7, 2013, www.vancouver foodpolicycouncil.ca/vfpc-member-helen-spiegleman-time-metro -vancouver-forget-about-incineration-get-going-composting.

25. N. Souto, D. Knapp, and N. Seldman, "Extended Producer Responsibility in British Columbia: A Work at Risk," part 3 of *The EPR Trilogy*, 2011 (Berkeley, CA: Urban Ore, 2012).

26. C. Morawski, "Understanding Economic and Environmental Aspects of Single-Stream Collection Systems," Container Recycling Institute, CRI, December, 2009, accessed June 4, 2012, www.container-recycling .org/publications/.

27. C. Morawski, "Understanding Economic and Environmental Aspects of Single-Stream Collection Systems."

28. N. Souto, D. Knapp, and N. Seldman, "Extended Producer Responsibility in British Columbia: A Work at Risk," 2011.

29. N. Souto, D. Knapp, and N. Seldman, "Extended Producer Responsibility in British Columbia: A Work at Risk," 2011.

30. "Zero Waste," Regional District of Nanaimo (website), accessed April 7, 2013, www.rdn.bc.ca/cms.asp?wpID=1063.

31. "Zero Waste," Regional District of Nanaimo (website).

32. "Zero Waste," Regional District of Nanaimo (website).

33. B. Boyd, personal communication, December 30, 2011. Buddy Boyd can be contacted at recycle@Gibsonsrecycling.ca, or visit him at www.gibsonsrecycling.ca.

34. G. Baldwin and J. Rustemeyer (producers), *Gibsons Resource Recovery Center*, DVD (Vancouver, BC: Peg Leg Films, 2012), accessed May 10, 2013, http://www.youtube.com/watch?v=t3Ut9gipa3M&list =UU7sLiFXXvPuzxXKV_sQpHwA&index=10&noredirect=1.

35. G. Baldwin and J. Rustemeyer (producers), *The Clean Bin Project* (Vancouver, BC: Peg Leg Films, 2010), accessed April 7, 2013, http:// cleanbinproject.com/the-film/.

36. B. Amundson, "Garbage Mining: Tapping into the Waste Revenue Stream," *The Dominion* (British Columbia), October 7, 2011, accessed April 7, 2013, www.dominionpaper.ca/articles/4179.

37. B. Amundson, "Garbage Mining: Tapping into the Waste Revenue Stream."

Chapter 10. Beyond Italy: Other Initiatives in Europe

1. C. Allen, "Flanders, Belgium: Europe's Best Recycling and Prevention Program," in C. Allen, V. Gokaldas, A. Larracas, et al., *On the Road to Zero Waste: Successes and Lessons from Around the World* (Berkeley, CA: Global Alliance for Incinerator Alternatives, 2012).

2. C. Allen, "Flanders, Belgium: Europe's Best Recycling and Prevention Program."

3. C. Allen, "Hernani, Spain: Door-to-Door Collection as a Strategy to Reduce Waste Disposal," in *On the Road to Zero Waste: Successes and Lessons from Around the World*.

4. C. Allen, "Hernani, Spain: Door-to-Door Collection as a Strategy to Reduce Waste Disposal."

5. R. Bailey and P. Connett (producers), *Community Composting in Zurich* (Canton, NY: VideoActive Productions, 1991). For copies contact Paul Connett, 104 Walnut Street, Binghamton, NY 13905, or pconnett@gmail.com.

6. C. Allen, "Hernani, Spain: Door-to-Door Collection as a Strategy to Reduce Waste Disposal."

7. Europa Zero Zabor, conference in San Sebastian, Spain, May 11–13, 2012.

8. Joan March Simon, personal communication, June 24, 2012. Simon can be reached by e-mail at jm.simon@no-burn.org.

9. Paul Martensson, personal communication, November 6, 2011. Martensson can be reached by e-mail at pal.martensson@kretslopp. goteborg.se and by mail at Pål Mårtensson, Department of Sustainable Water and Waste Management, Postgatan 16, Box 11192, SE-404 24 Göteborg, Sweden.

10. "Anger over Incinerator Decision in County," This is Gluocester (website), March 18, 2011, accessed April 7, 2013, www.thisis gloucestershire.co.uk/Anger-incinerator-decision-county/story -11890782-detail/story.html.

11. R. Bailey and P. Connett (producers), *Recycling in Germany* (Canton, NY: VideoActive Productions, 1986). For copies contact Paul Connett, 104 Walnut Street, Binghamton, NY 13905, or pconnett@gmail.com.

12. UK National Statistics, *Topic Guide to: Waste and Recycling 2012*, accessed April 7, 2013, www.statistics.gov.uk/hub/agriculture -environment/environment/waste-and-recycling.

13. J. McLaughlin. Personal communication, May 15, 2013. More information on Refurnish can be obtained at www.refurnish.co.uk, accessed May 15, 2013.

14. For the latest information on the Coventry Zero Waste Research Center please contact Jane Green (Friends of the Earth, UK) at janegreen@hotmail.com.

15. K. Anderson, personal communication, November 29, 2011. More information on the Cwm Harry program can be found at www .cwmharrylandtrust.org.uk/. Katy Anderson can be contacted at katy@cwmharry.org.uk.

16. More information on the Romaquip Kerb-Sort vehicle can be found at www.kerb-sort.com/.

17. "'Kerbsort': New Recycling Vehicle in Ballymena," *Ballymena Times*, November 11, 2011, accessed April 7, 2013, www.ballymenatimes .com/news/local/kerbsort_new_recycling_vehicle_in_ballymena _1_3208704.

18. "Municipal Sector Plan Part 1: Towards Zero Waste [WAG10-11169]," Welsh Assembly Government (website), March 22, 2011, accessed April 7, 2013, http://wales.gov.uk/docs/desh/publications /110310municipalwasteplan1en.pdf.

19. J. M. Simon, adapted from a power point presentation, "What EU Policies Prevent Zero Waste in the European Union?" given at a zero waste conference in Florence, Italy, September 17, 2012.

20. "Municipal Sector Plan Part 1: Towards Zero Waste [WAG10-11169]," Welsh Assembly Government (website), March 22, 2011, accessed April 7, 2013, http://wales.gov.uk/docs/desh/publications /110310municipalwasteplan1en.pdf.

21. "South Wales Incinerator: Viridor Cardiff Site Chosen," BBC News, February 1, 2013, accessed April 4, 2013, www.bbc.co.uk/news/uk -wales-21284964.

22. S. Gaskell, "Health Impact of Incinerators Must Be Studied, Says New Report," *Wales Online*, December 6, 2012, accessed April 4, 2013, www.walesonline.co.uk/news/wales-news/2012/12/06/health-impact -of-incinerators-must-be-studied-says-new-report-91466-32368938/.

Chapter 11. India, Japan, the Philippines, and Taiwan

1. K. K. Sruthijith, "Jindal Group's Upcoming Waste-to-Energy Plant Has Delhi Fuming," *ET Bureau*, October 6, 2011.
2. E. Balkan, "Government, Backed into a Corner on Public Incinerator Concerns, Pushes Back," *New Energy and Environment Digest*, November 12, 2009, accessed April 7, 2013, http://needigest.com /2009/11/12/government-backed-into-a-corner-on-public -incinerator-concerns-pushes-back/.
3. N. Tangri, "Respect for Recyclers: Protecting the Climate through Zero Waste," GAIA, October 2010, http://www.noburn.org /downloads/Respect%20for%20Recyclers%20(English)_1.pdf.
4. N. Tangri, personal communication, November 27, 2011. Tangri can be contacted at Neil@no-burn.org.
5. N. Tangri, "Pune, India: Waste Pickers Lead the Way to Zero Waste," in C. Allen, V. Gokaldas, A. Larracas, et al., *On the Road to Zero Waste: Successes and Lessons from Around the World* (Berkeley, CA: Global Alliance for Incinerator Alternatives, 2012).
6. N. Tangri, "Pune, India: Waste Pickers Lead the Way to Zero Waste."
7. V. Gokaldas, "Mumbai, India: Waste Picker–Run Biogas Plants as a Decentralized Solution," in C. Allen, V. Gokaldas, A. Larracas, et al., *On the Road to Zero Waste: Successes and Lessons from Around the World* (Berkeley, CA: Global Alliance for Incinerator Alternatives, 2012).
8. S. K. Nair and C. Jayakumar, *A Handbook on Waste Management for Rural Tourism Areas: A Zero Waste Approach* (Geneva: United Nations Development Program India, 2008), accessed May 10, 2013, www .undp.org.in. For more information on this project see the website of the group Thanal at www.thanal.co.in/index.php?option=com_content &view=article&id=61&Itemid=94. Jayakumar can be reached at jayakumar.c@gmail.com.
9. "Zero Waste in Kamikatsu, Japan," BBC News, July 11, 2008, accessed April 7, 2013, http://news.bbc.co.uk/2/hi/7499954.stm.
10. J. T. Gamil and K. L. Alave, "Green Groups Block Return of Incinerators," *Philippine Daily Inquirer*, September 5, 2011, accessed April 7, 2013, http://newsinfo.inquirer.net/53571/%E2%80%98green %E2%80%99-groups-block-return-of-incinerators.
11. F. Grate, personal communication, May 10, 2013. Grate can be reached at froilan@motherearthphil.org.
12. F. Grate, personal communication, June 12, 2012.

13. F. Grate, personal communication, June 20, 2012.

14. S. Mendoza, "Citywide Plastic Ban: Muntinlupa Takes Giant Step," *Philippine Daily Inquirer*, January 29, 2011, accessed April 7, 2013, http://opinion.inquirer.net/inquireropinion/talkofthetown/view /20110129-317364/Muntinlupa-takes-giant-step.

15. "Should Plastic Bags Be Banned?" Debatewise (website), accessed April 7, 2013, http://debatewise.org/debates/1011-should-plastic -bags-be-banned.

16. G. Chen and H. Hsieh, "The Problems of Incinerators in Taiwan," 2011, accessed January 20, 2011, www.taiwanwatch.org.tw/Twweb _english/index.htm.

17. Taiwan Review, "Recycling: Taiwan's Way of Life," March 1, 2010, an official publication of the Taiwan government, accessed May 13, 2013, http://taiwanreview.nat.gov.tw/ct.asp?xItem=93434&CtNode=1337.

Chapter 12. Bahrain, Egypt, and Lebanon

1. "Bahrain to Work 'Towards Zero Waste,'" *Trade Arabia*, November 15, 2011, accessed May 13, 2013, www.tradearabia.com/news/ENV _208013.html.

2. E. Wassef (director), *Marina of the Zabbaleen* (New York: Torch Films, 2008).

3. M. Iskander (director), *Garbage Dreams* (New York: Iskander, 2009).

4. J. Kramer (director), *Zabbaleen: A Documentary Film* (2012) accessed July 28, 2013, www.everydaytrash.com/2012/07/20/zabaeleen-the-movie/.

5. G. Cruz, personal communication, January 20, 2011.

6. C. Brady (director), *Trashed*, an environmental documentary feature film (Watlington, Oxfordshire, UK: Blenheim Films, 2012), accessed April 7, 2013, www.trashedfilm.com/about-trashed.php.

Chapter 13. South America

1. J. Gutberlet, *Recovering Resources, Recycling Citizenship: Urban Poverty Reduction in Latin America* (Aldershot: Ashgate, 2008).

2. J. Gutberlet, *Recovering Resources, Recycling Citizenship: Urban Poverty Reduction in Latin America*, 125.

3. E. F. Schumacher, *Small Is Beautiful: Economics as if People Mattered* (London: Blond and Briggs, 1973).

4. J. Gutberlet, *Recovering Resources, Recycling Citizenship: Urban Poverty Reduction in Latin America*, 148.

5. J. Gutberlet, *Recovering Resources, Recycling Citizenship: Urban Poverty Reduction in Latin America*, 4.

6. J. Gutberlet, "Informal and Cooperative Recycling as a Poverty Eradication Strategy," *Geography Compass*, 6 (2012): 19–34. The specific quotes were obtained from a poster presentation, J. Gutberlet, "Inclusive Solid Waste Management: Urban Adaptation Towards Greater Resource Efficiency."

7. C. Allen, "Buenos Aires City, Argentina: Including Grassroots Recyclers," in C. Allen, V. Gokaldas, A. Larracas, et al., *On the Road to Zero Waste: Successes and Lessons from Around the World* (Berkeley, CA: Global Alliance for Incinerator Alternatives, 2012).

8. C. Allen, "Environmental Possibilities: Vegetable Waste to Zero Waste in La Pinata, Chile," in C. Allen, V. Gokaldas, A. Larracas, et al., *On the Road to Zero Waste: Successes and Lessons from Around the World* (Berkeley, CA: Global Alliance for Incinerator Alternatives, 2012).

9. S. M. Dias, "Integrating Informal Workers into Selective Waste Collection: The Case of Belo Horizonte, Brazil," Women in Informal Employment: Globalizing and Organizing, Urban Policies Briefing Note No. 6, 2011, accessed May 10, 2013, http://wiego.org/sites/wiego.org/files/publications/files/Dias_WIEGO_PB4.pdf.

10. S. M. Dias, "The Municipal Waste and Citizenship Forum: A Platform for Social Inclusion and Participation," Women in Informal Employment: Globalizing and Organizing, Urban Policies Briefing Note No. 5, 2011, accessed May 10, 2013, http://wiego.org/sites/wiego.org/files/publications/files/Dias_WIEGO_PB5.pdf.

11. S. M. Dias, "Overview of the Legal Framework for Inclusion of Informal Recyclers in Solid Waste Management in Brazil," Women in Informal Employment: Globalizing and Organizing, Urban Policies Briefing Note No. 6, 2011, accessed May 10, 2013, http://wiego.org/sites/wiego.org/files/publications/files/Dias_WIEGO_PB6.pdf.

12. S. M. Dias, "Trajetórias e Memórias dos Fóruns Lixo e Cidadania no Brasil: Experimentos Singulares de Justiça Social e Governança Participativa" [Trajectories and memories of the waste and citizenship forums: unique experiments of social justice and participatory governance], PhD thesis, State University of Minas Gerais, 2009.

13. S. M. Dias, J. Ijgosse, and R. T. V. Barros, "Belo Horizonte City Profile," in UN Habitat, *Solid Waste Management, Water, and Sanitation in the World's Cities* (London: Earthscan, 2010).

14. S. M. Dias and F. C. G. Alves, "Integration of the Informal Recycling Sector in Solid Waste Management in Brazil," 2008, accessed May 10, 2013, http://www.giz.de/Themen/en/dokumente/gtz2010-waste -experts-conditions-is-integration.pdf.

15. S. M. Dias, "Building Citizenship: Achievements and Limitations of the Selective Waste Collection Project in Partnership with ASMARE" (master's dissertation, Federal University of Minas Gerais State, 2002).

17. S. M. Dias, "Integrating Waste Pickers/Catadores for Sustainable Recycling," paper presented at the workshop Planning for Sustainable and Integrated Solid Waste Management, SKAT, Manila, Philippines, September 2000, accessed May 10, 2013, www.cwgnet.

16. M. Donoso, personal communication, August 19, 2011. Donoso can be contacted at magdalena@no-burn.org.

18. Bulletin of the Global Alliance of Waste Pickers (GAWP), December 21, 2011, accessed May 10, 2013, www.globalrec.org/2011/12/21 /a-huge-victory-for-waste-pickers-in-colombia/.

19. S. M. Dias, "Recycling in Belo Horizonte—An Overview of Inclusive Programming," Women in Informal Employment: Globalizing and Organizing, Urban Policies Briefing Note No. 5, 2011, accessed May 10, 2013, http://wiego.org/publications/recycling-belo-horizonte-brazil ---overview-inclusive-programming.

Chapter 14. Zero Waste and the Local Economy

1. The Institute for Local Self-Reliance (ILSR) has a website at www.ilsr.org.

2. N. Seldman, "History of the US Recycling Movement," in *Encyclopedia of Energy Technology and the Environment*, eds. A. Bistro and S. Boots (New York: Wiley Brothers, 1995).

3. T. Allan, B. Platt, and D. J. Morris, *Beyond 25 Percent: Materials Recovery Comes of Age* (Washington, DC: Institute for Local Self-Reliance, 1989).

4. B. Platt, C. Doherty, A. C. Broughton, et al., *Beyond 40 Percent: Record-Setting Recycling and Composting Programs* (Washington, DC: Institute for Local Self-Reliance, 1991).

5. B. Platt, N. Friedman, C. Grodinsky, et al., *In-Depth Studies of Recycling and Composting Programs: Designs, Costs, Results*, 3 vols. (Washington, DC: Institute for Local Self-Reliance, 1992).

6. B. Platt and K. Lease, *Cutting the Waste Stream in Half: Community Record-Setters Show How* (Washington, DC: Institute for Local Self-Reliance, 1999).

7. B. Platt and N. Seldman, *Wasting and Recycling in the US 2000* (Athens, GA: GrassRoots Recycling Network, 2000).

8. A complete listing of the incinerators defeated in the United States between 1985 and 1995 can be found in *Waste Not* issues #251–74 (1994). Back issues of *Waste Not* are available at www.americanhealthstudies.org.

9. R. W. Beck, *U.S. Recycling Economic Information Study* (Washington, DC: National Recycling Coalition, 2001), accessed May 10, 2013, www.epa.gov/osw/conserve/rrr/rmd/rei-rw/pdf/n_report.pdf.

10. Greenaction and Global Alliance for Incinerator Alternatives, *Incinerators in Disguise: Case Studies of Gasification, Pyrolysis, and Plasma in Europe, Asia, and the United States* (Berkeley, CA: Global Alliance for Incinerator Alternatives, 2006), accessed April 1, 2013, www.greenaction .org/incinerators/documents/IncineratorsInDisguiseReportJune2006.pdf.

11. J. Ostrowski, "Selection Committee Picks Low Bidder to Build West Palm Beach Incinerator Plant," *Palm Beach Post*, March 16, 2011, accessed May 10, 2013, www.palmbeachpost.com/money/selection -committee-picks-low-bidder-to-build-west-1326353.html.

12. S. Cointreau, J. M. Huls, N. Seldman, et al., *Municipal Waste Management and Recycling* (Geneva: World Bank, 1980).

13. The GrassRoots and Recycling Network has a website at www.grrn.org/.

14. The ACT no waste law was passed in 1996. The website for the administration of this law is www.tams.act.gov.au/live/recycling-waste /about_ACT_NOWaste.

15. P. Connett (producer), *Zero Waste: Idealistic Dream or Realistic Goal?* (Grass Roots and Global Video, 1999), accessed May 10, 2013, www.americanhealthstudies.org/video.html#waste.

16. The Zero Waste International Alliance definition of zero waste can be accessed at http://zwia.org/joomla/index.php?option=com_content &view=article&id=9&Itemid=6.

Chapter 15. Waste Isn't Waste Until It's Wasted

1. R. W. Beck, *U.S. Recycling Economic Information Study* (Washington, DC: National Recycling Coalition, 2001), accessed May 10, 2013, www.epa.gov/osw/conserve/rrr/rmd/rei-rw/pdf/n_report.pdf.

2. P. Connett (producer), *Zanker Road: A Different Kind of Landfill* (Canton, NY: Grass Roots and Global Video, 2000). For copies contact Paul Connett, 104 Walnut Street, Binghamton, NY 13905, or pconnett@gmail.com.

Chapter 16. Waste Is a Social Issue First, a Market Issue Second

1. A. Underwood, "10 Fixes for the Planet," *Newsweek*, April 14, 2008, accessed May 10, 2013, www.thedailybeast.com/newsweek/2008/04/05 /10-fixes-for-the-planet.html.

Chapter 17. Gibsons Resource Recovery Center

1. The Recycling Council of British Columbia (RCBC) has a website at http://rcbc.bc.ca/.
2. The Zero Waste International Alliance (ZWIA) has a website at www.zwia.org.
3. G. Baldwin and J. Rustemeyer (producers), *Gibsons Resource Recovery Center*, DVD (Vancouver, BC: Peg Leg Films, 2012), accessed April 7, 2013, www.youtube.com/user/GibsonsRecycling1?feature=mhum#p/a /u/0/t3Ut9gipa3M.
4. G. Baldwin and J. Rustemeyer (producers), *Clean Bin Project*, DVD (Vancouver, BC: Peg Leg Films, 2012), accessed April 7, 2013, http:// cleanbinproject.com/the-film/.
5. "The Zero Waste Challenge," Metro Vancouver (website), accessed May 10, 2013, www.metrovancouver.org/services/solidwaste/Residents /zerowaste/Pages/default.aspx.
6. "The Zero Waste Challenge Conference," Metro Vancouver (website), accessed May 10, 2013, www.metrovancouver.org/REGION /ZEROWASTECONFERENCE/Pages/default.aspx.
7. F. Bula, "Incineration Firms Vying to Burn Vancouver's Trash," *Globe and Mail* (Vancouver), January 3, 2012, accessed May 10, 2013, http:// www.theglobeandmail.com/news/british-columbia/incineration -firms-vying-to-burn-vancouvers-trash/article4097966/.
8. "Zero Waste in Kamikatsu, Japan," BBC News, July 11, 2008, accessed April 7, 2013, http://news.bbc.co.uk/2/hi/7499954.stm.

Chapter 18. Multimaterial Curbside Recycling and Producer Responsibility

1. The Campaign for Real Recycling (Canada) has a website at www. campaignforrealrecycling.ca/; the Campaign for Real Recycling (UK) has a website at www.realrecycling.org.uk/.
2. S. MacBride, *Recycling Reconsidered: The Present Failure and Future Promise of Environmental Action in the United States* (Cambridge: MIT Press, 2012).

3. H. Crooks, *Giants of Garbage: The Rise of the Global Waste Industry and the Politics of Pollution Control* (Toronto: Lorimer, 1993). Crooks adopted the phrase "municipal industrial complex" to describe this working relationship, a phrase that obviously echoes the relationship of governments and the weapons industry.

4. M. V. Melosi, *Garbage in the Cities: Refuse, Reform and the Environment* (College Station: Texas A&M University Press, 1981).

5. J. Stolaroff, *Products, Packaging and US Greenhouse Gas Emissions* (Athens, GA: Product Policy Institute, 2009).

6. C. Rootes and L. Leonard, "Environmental Movements and Campaigns Against Waste Infrastructure in the United States," *Environmental Politics*, 18(6) (2009), accessed June 4, 2012, www .kent.ac.uk/sspssr/staff/academic/rootes/usa-waste.pdf.

7. C. Morawski, "Understanding Economic and Environmental Aspects of Single-Stream Collection Systems," Culver City, CA: Container Recycling Institute (CRI), 2009, accessed June 4, 2012, www.container -recycling.org/publications/.

8. Product Policy Institute, "Detailed Calculation for Preliminary Calculation of Blue Box Steward Fees," August 2009, accessed June 7, 2012, www.productpolicy.org/ppi/2010-Stewardship-Ontario_Blue -Box_Tables-1and2.xls.

9. Container Recycling Institute (CRI), "Single-Stream Recycling," CRI website, 2012, accessed May 10, 2013, http://www.container-recycling .org/index.php/publications.

10. D. Menzies, "Waste Blues: Curbside Recycling Reassessed," *The Financial Post Magazine*, September 1997, accessed June 4, 2012, www.web.ca/~jjackson/wstblues.html.

11. G. Crittenden, "Discussion Group:The Blue Box Conspiracy," *Next City*, September 21, 1997, accessed May 10, 2013, http:// urbanrenaissance.probeinternational.org/1997/09/21/discussion -group-blue-box-conspiracy/.

12. Vermont Bill as Introduced, H.696 (2010), www.leg.state.vt.us/docs /2010/bills/intro/H-696.pdf.

13. "Recycling Reinvented, 2012: Frequently Asked Questions," Recycling Reinvented website, accessed June 4, 2012, http://recycling-reinvented .org/?page_id=99.

14. Elizabeth Royte blog, "Bottle Water Industry and RFK Jr. Sue New York State," May 26, 2009, accessed June 8, 2012, www.royte.com/blog/?p=333.

15. Guy Crittenden, "Discussion Group: The Blue Box Conspiracy," August 7, 2009, http://eprf.probeinternational.org/node/1976.

16. Tellus Institute with Sound Resource Management Group, *More Jobs, Less Pollution: Growing the Recycling Economy in the U.S.*, prepared for the Blue Green Alliance, www.bluegreenalliance.org, accessed May 10, 2013, www.container-recycling.org/assets/pdfs/jobs/MoreJobs LessPollution.pdf.

17. J. Morris and C. Morawski, *Returning to Work: Understanding the Domestic Jobs Impacts from Different Methods of Recycling Beverage Containers* (Culver City, CA: Container Recycling Institute, 2011), accessed May 10, 2013, www.container-recycling.org/assets/pdfs /reports/2011-ReturningToWork.pdf.

18. "Metro Vancouver Waste Plan," Metro Vancouver, accessed June 11, 2012, www.metrovancouver.org/services/solidwaste/planning/Pages /default.aspx, see page 26. Metro Vancouver often cites the example of Europe, which relies heavily on waste incineration. Under the EU Packaging Directive as amended in 2003, energy recovery through incineration was accorded equal status with recycling as an outcome for packaging materials collected in EPR programs. However, this policy has been effectively reversed by a recent decision by the European Commission to eliminate burning of recyclable and compostable materials in incinerators in the future.

19. "Less Garbage, or End to Recycling?" *Vancouver Sun*, April 9, 2012, accessed June 11, 2012, www.vancouversun.com/news/Less+garbage +recycling/6429601/story.html.

Chapter 19. Producer Responsibility, the Cornerstone of Zero Waste

1. The GrassRoots and Recycling Network (GRRN) has a website at www.grrn.org/.

2. P. Connett (producer), *Zero Waste: Idealistic Dream or Realistic Goal?* (Canton, NY: Grass Roots and Global Video, 1999), accessed May 10, 2013, www.americanhealthstudies.org/video.html#waste.

3. P. Connett and B. Sheehan, *A Citizens' Agenda for Zero Waste: A United States/Canadian Perspective* (Athens, GA: GrassRoots and Recycling Network, 2001), accessed May 10, 2013, http://archive.grrn.org /zerowaste/community/activist/citizens_adenda_4_print.pdf.

4. The Product Policy Institute (PPI) has a website at www.productpolicy.org/.

5. Canadian Council of Ministers for the Environment (CCME), *Canada-wide Action Plan for Extended Producer Responsibility,* October 29, 2009, accessed May 10, 2013, www.ccme.ca/assets/pdf/epr_cap.pdf

6. British Columbia Ministry of the Environment, *Product Stewardship in British Columbia,* 2011, accessed, May 10, 2013, www.env.gov.bc.ca /epd/recycling/history/index.htm.

7. The Electronics TakeBack Coalition has a website at www.electronicstakeback.com/home/.

8. The stated mission of the Container Recycling Institute (CRI) is "to make North America a global model for the collection and quality recycling of packaging materials. We envision a world where no material is wasted, and the environment is protected. We succeed because companies and people collaborate to create a strong, sustainable domestic economy." The CRI has a website at www.container -recycling.org/.

9. The Mercury Policy Project has the mission of "promoting policies to eliminate mercury use and reduce mercury exposure." The project has a website at http://mercurypolicy.org/.

10. The California Product Stewardship Council has a website at www.calpsc.org/.

11. The Northwest Product Stewardship Council has a website at www.productstewardship.net/.

12. The New York Product Stewardship Council has a website at www.nypsc.org/.

13. The Texas Product Stewardship Council has a website at www.txpsc.org/.

14. The Vermont Product Stewardship Council has a website at www.vtpsc.org/.

15. For information on product stewardship and extended producer responsibility go to the PPI website at www.productpolicy.org /content/epr-policy.

16. For information on EPR resolutions in six states go to the PPI website at www.productpolicy.org/content/adopt-epr-resolutions.

17. For information on national EPR resolutions go to the PPI website at www.productpolicy.org/content/national-epr-resolutions.

18. Cradle2 is a takeoff on the notion that industry should take responsibility "cradle to cradle" for it products. The website for Cradle2 is www.cradle2.org.

19. For information on EPR laws go to the PPI website at www.productpolicy.org/content/epr-laws.

20. For information on framework EPR legislation go to the PPI website at www.productpolicy.org/content/framework-north-america.

21. For information on the EPR starter kit go to the PPI website at www.productpolicy.org/ppi/PPI_EPR_Starter_Kit_.pdf.

22. Recycling Council of British Columbia (RCBC), "Packaging and Printer Paper EPR in British Columbia," Vancouver, BC; RCBC, 2011, accessed May 10, 2013, http://rcbc.bc.ca/education/product -stewardship/#packaging.

23. B. Kyle, "E-Waste Export Legislation Is the Most Important Action the Federal Government Can Take on E-Waste Problems," Electronics TakeBack Coalition, June 23, 2011, accessed May 10, 2013, www .electronicstakeback.com/2011/06/23/e-waste-export-legislation/.

24. "First State Producer Responsibility 'Framework' Law Passed in Maine with Unanimous Bi-Partisan and Chamber of Commerce Support," Product Policy Institute, March 23, 2010, accessed May 10, 2013, www.productpolicy.org/ppi-press-release/first-state -producer-responsibility-framework-law-passed-maine -unanimous-bi-partis.

25. See information on EPR bills passed in California at the California Product Stewardsh ip Council website at www.calpsc.org/.

26. Product Management Alliance (PMA), "The Product Management Alliance Launched to Promote and Protect Free-Market Product Stewardship Solutions," press release, September 27, 2011, accessed May 10, 2013, http://productmanagementalliance.org/.

Chapter 21. The Economics of Zero Waste

1. The Sound Resource Management Group has a website at www.zerowaste.com.

2. Earth Track states its purpose as "understanding and monitoring environmentally-harmful subsidies" and has a website at www.earthtrack.net.

3. Tellus Institute with Sound Resource Management Group, "More Jobs, Less Pollution: Growing the Recycling Economy in the U.S.," prepared for the Blue Green Alliance, www.bluegreenalliance.org, accessed May 10, 2013, www.container-recycling.org/assets/pdfs /jobs/MoreJobsLessPollution.pdf.

4. J. Morris and C. Morawski, *Returning to Work: Understanding the Domestic Jobs Impacts from Different Methods of Recycling Beverage Containers* (Culver City, CA: Container Recycling Institute, 2011), accessed May 10, 2013, www.container-recycling.org/assets/pdfs /reports/2011-ReturningToWork.pdf.

Chapter 22. Businesses Are Leading the Way to Zero Waste

1. P. Connett (producer), *Zero Waste: Idealistic Dream or Realistic Goal?* (Canton, NY: Grass Roots and Global Video, 1999), accessed May 10, 2013, www.americanhealthstudies.org/video.html#waste.
2. See the GrassRoots and Recycling Network's ten zero waste business principles at www.grrn.org/page/zero-waste-business-principles.
3. See the Zero Waste International Alliance zero waste business principles at http://zwia.org/joomla/index.php?option=com_content &view=article&id=8&Itemid=7.
4. See PowerPoints from zero waste businesses telling their own story about how they achieved zero waste at www.earthresource.org/zerowaste.html.
5. I. Amato, "Can We Make Garbage Disappear?" *Time,* November 8, 1999, accessed May 10, 2013, www.time.com/time/magazine/article /0,9171,992527,00.html.
6. A. Paulson, "So You Think Your'e Recycling at Work?" *Christian Science Monitor,* September 10, 2001, accessed May 10, 2013, www.csmonitor .com/2001/0910/p11s1-wmwo.html.
7. M. Gunther, "The End of Garbage," *Fortune,* March 19, 2007, accessed May 10, 2013, http://money.cnn.com/magazines/fortune/fortune _archive/2007/03/19/8402369/index.htm.
8. C. Woodyard, "It's Waste Not, Want Not at Super Green Subaru Plant," *USA Today,* February 19, 2008, accessed May 10, 2013, www.usatoday .com/money/industries/environment/2008-02-18-green-factories_N.htm.
9. A. Underwood, "10 Fixes for the Planet," *Newsweek,* April 14, 2008, accessed May 10, 2013, http://www.thedailybeast.com/newsweek/2008 /04/05/10-fixes-for-the-planet.html.
10. D. Ferry, "The Urban Quest for 'Zero' Waste," *Fortune,* September 12, 2011, accessed May 10, 2013, http://online.wsj.com/article/SB1000142 4053111904583204576542233226922972.html.
11. See the Zero Waste International Alliance definition of *zero waste* at http://zwia.org/joomla/index.php?option=com_content&view=article &id=12&Itemid=5.

Chapter 23. The Zero Waste International Alliance

1. The Californian Resource Recovery Association (CRRA) was founded in 1974 and is the oldest and one of the largest nonprofit recycling organizations in the United States; it has a website at www.crra.com.

2. The GrassRoots and Recycling Network (GRRN) has a website at www.grrn.org.

3. The Zero Waste International Alliance (ZWIA) has a website at www.zwia.org.

4. The Zero Waste San Diego organization has a website at http://zerowastesandiego.org/.

5. P. Connett (producer), *Zero Waste: Idealistic Dream or Realistic Goal?* (Canton, NY: Grass Roots and Global Video, 1999), accessed May 10, 2013, www.americanhealthstudies.org/video.html#waste.

6. EMPA. According to Wikipedia, EMPA is the German acronym for Eidgenössische Materialprüfungs und Forschungsanstalt. In English this translates to the "Swiss Federal Laboratories for Materials Science and Technology." Wikipedia goes on to describe EMPA as "an interdisciplinary Swiss research and service institution for applied materials sciences and technology. As part of the Swiss Federal Institutes of Technology Domain, it is an institution of the Swiss federation. For the longest time since its foundation in 1880, it concentrated on classical materials testing. Since the late 1980s, it has developed into a modern research and development institute."

7. Communities Against Toxics (CATS), under the leadership of Ralph Ryder, is a grassroots group in the UK, which for over twenty years has been fighting incinerators and other polluting projects in the UK and beyond. The group circulates a monthly newsletter. Its website is http://www.communities-against-toxics.org.uk/.

8. R. Murray, *Creating Wealth from Waste* (London: Demos, 1999).

9. R. Murray, *Zero Waste* (London: Greenpeace, 2002). More information about this book can be accessed at www.greenpeace.org .uk/media/press-releases/britain-could-be-a-rubbish-free-society -says-ground-breaking-study.

RESOURCES

Books and Articles

Louis Blumberg and Robert Gottlieb, *War on Waste: Can America Win Its Battle with Garbage?* (Washington, DC: Island Press, 1989).

Rachel Carson, *Silent Spring* (New York: Houghton Mifflin, 1962).

Thomas H. Christensen (editor), *Solid Waste Technology and Management*, 2 vols. (New York: Wiley, 2010).

Theo Colborn, Dianne Dumanoski, and John Peterson Myers, *Our Stolen Future: Are We Threatening Our Fertility, Intelligence, and Survival? A Scientific Detective Story* (New York: Dutton, 1996).

Barry Commoner, *The Closing Circle* (New York: Knopf, 1971).

Barry Commoner, *Making Peace with the Planet* (New York: Pantheon, 1990).

Paul Connett, Rossano Ercolini, and Patrizia Lo Sciuto, *Rifiuti Zero: Una Rivoluzione in Corso* (Viareggio, Italy: Disessensi, 2012).

Jennifer Corson, *The Resourceful Renovator* (Toronto: Key Porter Books, 2000).

Harold Crooks, *Giants of Garbage: The Rise of the Global Waste Industry and the Politics of Pollution Control* (Toronto: Lorimer, 1993).

Alan T. Durning, *How Much Is Enough? The Consumer Society and the Future of the Earth* (New York: W. W. Norton, 1992).

Lois M. Gibbs, *Dying from Dioxin: A Citizen's Guide to Reclaiming Our Health and Rebuilding Democracy* (Boston: South End Press, 1995).

Paul Hawken, *The Ecology of Commerce: A Declaration of Sustainability* (New York: HarperCollins, 1995); see also the revised edition (2010).

David Korten, *When Corporations Rule the World* (Boulder, CO: Kumarian, 1995).

Annie Leonard, *The Story of Stuff: How Our Obsession with Stuff Is Trashing the Planet, Our Communities, and Our Health—and a Vision for Change* (New York: Free Press, 2010).

Samantha MacBride, *Recycling Reconsidered: The Present Failure and Future Promise of Environmental Action in the United States* (Cambridge, MA: MIT Press, 2012).

Jerry Mander, *Four Arguments for the Elimination of Television* (New York: HarperCollins, 1978).

Jerry Mander, *In the Absence of the Sacred: The Failure of Technology and the Survival of the Indian Nations* (San Francisco: Sierra Club, 1992).

William McDonough and Michael Braungart, *Cradle to Cradle: Remaking the Way We Make Things* (New York: Northpoint, 2002).

Donella H. Meadows, Dennis L. Meadows, Jorgen Randers, and William W. Behrens III, *The Limits to Growth* (Washington, DC: Potomac Associates, 1972).

Martin V. Melosi, *Garbage in the Cities: Refuse, Reform and the Environment* (College Station: Texas A&M University Press, 1981).

Robin Murray, *Creating Wealth from Waste* (London: Demos, 1999).

Robin Murray, *Zero Waste* (London: Greenpeace, 2002).

Newsday, *Rush to Burn: Solving America's Garbage Crisis?* (Washington, DC: Island Press, 1988).

Vance Packard, *The Waste Makers* (New York: D. McKay, 1960).

Brenda Platt and Neil Seldman, *Wasting and Recycling in the United States 2000* (Athens, GA: GrassRoots Recycling Network, 2000).

Jorgen Randers, *2052: A Global Forecast for the Next Forty Years* (White River Junction, VT: Chelsea Green, 2012).

Ernst Friedrich Schumacher, *Small Is Beautiful: Economics as if People Mattered* (London: Blond and Briggs, 1973).

Neil Seldman, "History of the US Recycling Movement," in *Encyclopedia of Energy Technology and the Environment*, ed. A. Bistro and S. Boots (New York: Wiley Brothers, 1995).

Joshuah Stolaroff, *Products, Packaging and US Greenhouse Gas Emissions* (Athens, GA: Product Policy Institute, 2009).

Joel Tickner, Carolyn Raffensperger, and Nancy Myers, "The Precautionary Principle in Action: A Handbook," Science and Environmental Health Network, accessed July 29, 2012, www.mindfully.org/Precaution /Precaution-In-Action-Handbook.pdf.

Barbara Ward and Rene Dubos, *Only One Earth* (New York: W. W. Norton, 1972).

Reports

T. Allan, B. Platt, and D. J. Morris, *Beyond 25 Percent: Materials Recovery Comes of Age* (Washington, DC: Institute for Local Self-Reliance, 1989).

C. Allen, V. Gokaldas, A. Larracas, et al., *On the Road to Zero Waste: Successes and Lessons from Around the World* (Berkeley, CA: Global Alliance for Incinerator Alternatives, 2012), accessed April 2, 2013, www.no-burn .org/on-the-road-to-zero-waste-successes-and-lessons-from-around -the-world.

R. W. Beck, *U.S. Recycling Economic Information Study* (Washington, DC: National Recycling Coalition, 2001), accessed April 7, 2013, www.epa.gov /osw/conserve/rrr/rmd/rei-rw/pdf/n_report.pdf.

P. Connett and B. Sheehan, *A Citizen's Agenda for Zero Waste: A United States/ Canadian Perspective* (Athens, GA: GrassRoots Recycling Network, 2001), accessed May 10, 2013, http://archive.grrn.org/zerowaste /community/activist/citizens_adenda_4_print.pdf.

Eco-Cycle, *Zero Waste and Climate Change,* website discussion, accessed April 8, 2013, www.ecocycle.org/zerowaste/climate.

Eco-Cycle, Global Alliance for Incinerator Alternatives, and the Institute for Local Self-Reliance, *Stop Trashing the Planet,* June 5, 2008, accessed April 8, 2013, www.stoptrashingtheclimate.org/.

Environmental Protection Agency, *Municipal Solid Waste (MSW) in the United States: Facts and Figures,* 2010, accessed April 8, 2013, www.epa.gov/epawaste/nonhaz/municipal/msw99.htm.

Franklin Associates, *The Role of Recycling in Integrated Solid Waste Management to the Year 2000* (Stamford, CT: Keep America Beautiful, 1994).

GrassRoots Recycling Network, *Welfare for Waste: How Federal Taxpayer Subsidies Waste Resources and Discourage Recycling* (Athens, GA: GrassRoots Recycling Network, 1999), accessed April 9, 2013, www.grrn.org/assets/pdfs/wasting/w4w.pdf.

Greenaction and Global Alliance for Incinerator Alternatives, *Incinerators in Disguise: Case Studies of Gasification, Pyrolysis, and Plasma in Europe, Asia, and the United States* (Berkeley, CA: Global Alliance for Incinerator Alternatives, 2006), accessed April 1, 2013, www.greenaction.org /incinerators/documents/IncineratorsInDisguiseReportJune2006.pdf.

ICF Consulting, "Incineration of Municipal Solid Waste: A Reasonable Energy Option?" Fact Sheet #3 (2005), accessed April 1, 2013, www.durhamenvironmentwatch.org/Incinerator%20Files/FS3energy.pdf.

"Measuring Sustainable Development: Application of the Genuine Progress Index to Nova Scotia," Nova Scotia GPI Solid Waste-Resource Accounts, 2004, accessed May 14, 2013, www.gpiatlantic.org/pdf/solidwaste /solidwaste.pdf.

J. Morris, *Review of Waste Management Options,* report for the City of Halifax (Seattle, WA: Sound Resource Management Group, 1992).

J. Morris and D. Canzoneri, *Recycling versus Incineration: An Energy Conservation Analysis* (Seattle, WA: Sound Resource Management Group, 1993).

J. Morris and C. Morawski, *Returning to Work: Understanding the Domestic Jobs Impacts from Different Methods of Recycling Beverage Containers* (Culver City, CA: Container Recycling Institute, 2011), accessed May 10, 2013, www.container-recycling.org/assets/pdfs/reports/2011-ReturningToWork.pdf.

B. Platt, C. Doherty, A. C. Broughton, et al., *Beyond 40 Percent: Record-Setting Recycling and Composting Programs* (Washington, DC: Institute for Local Self-Reliance, 1991).

B. Platt, N. Friedman, C. Grodinsky, et al., *In-Depth Studies of Recycling and Composting Programs: Designs, Costs, Results*, 3 vols. (Washington, DC: Institute for Local Self-Reliance, 1992).

B. Platt and K. Lease, *Cutting the Waste Stream in Half: Community Record-Setters Show How* (Washington, DC: Institute for Local Self-Reliance, 1999).

N. Tangri, "Respect for Recyclers" (Berkeley, CA: GAIA, October 2010), accessed May 14, 2013, http://www.noburn.org/downloads/Respect%20for%20Recyclers%20(English)_1.pdf.

N. Tangri, *Waste Incineration: A Dying Technology* (Berkeley, CA: GAIA, July 14, 2003), accessed April 8, 2013, www.no-burn.org/article.php?id=276.

Tellus Institute with Sound Resource Management Group, *More Jobs, Less Pollution: Growing the Recycling Economy in the U.S.* Prepared for the Blue Green Alliance, www.bluegreenalliance.org, accessed May 10, 2013, www.container-recycling.org/assets/pdfs/jobs/MoreJobsLessPollution.pdf.

World Bank, *What a Waste: A Global Review of Solid Waste Management*, June 6, 2012, accessed May 14, 2013 http://web.worldbank.org/WBSITE/EXTERNAL/TOPICS/EXTURBANDEVELOPMENT/0,,contentMDK:23172887~pagePK:210058~piPK:210062~theSitePK:337178,00.html

Films

C. Brady (director), *Trashed* (Watlington, Oxfordshire, UK: Blenheim Films, 2012), accessed April 7, 2013, www.trashedfilm.com/about-trashed.php.

V. Ibanez (director), *Zero Waste* (Italy: Ibanez, 2011), accessed May 9, 2013, www.cinemambiente.it/film_ambiente/5598_.html.

M. Iskander (director), *Garbage Dreams* (Cairo, Egypt: Wyn Films, 2009), accessed May 14, 2013, www.firstshowing.net/.../sxsw-film-festival-2009-full-line-up-announced/.

Justin Kramer, *Zabbaleen: A Documentary Film*, still in production, accessed May 14, 2013, www.kickstarter.com/projects/justinkramer/zabaleen-a-documentary-film.

A. Leonard, *The Story of Stuff,* video accessed May 10, 2013, www.storyofstuff.com/.

Engi Wassef (director), *Marina of the Zabbaleen* (New York: Torch Films, 2008), accessed May 14, 2014, http://www.amazon.com/Marina -Zabbaleen-Engi-Wassef/dp/B002VB4CK0.

TV Programs

J. de Graaf and V. Boe (producers), *Affluenza* (KCTS/Seattle and Oregon Public Broadcasting, aired September 15,1997).

J. de Graaf and V. Boe (producers), *Escape from Affluenza,* 1998. Website accessed May 10, 2013, http://www.pbs.org/kcts/affluenza/.

M. Ostrow (director), *Race to Save the Planet* (Annenberg PBS, 1990).

Videos

Between 1985 and 1995 Paul Connett and Roger Bailey (of VideoActive Productions) made a series of videos on incinerators of different kinds (municipal, medical, and hazardous) as well as videos about individuals and communities pursuing more sustainable alternatives. The most relevant today is *Waste Management: As if the Future Mattered,* and this can be viewed online at www.americanhealthsciences.org.

A complete listing of the videotapes with descriptions, including the ten videotapes on dioxin shot at the First Citizens Conference on Dioxin (1991) can be found on the website www.americanhealthsciences.org. Several of these videos have been referenced in this text; they are listed below. All the videos have now been transferred to DVD and copies can be obtained by contacting Paul Connett at 104 Walnut Street, Binghamton, NY 13905, or pconnett@gmail.com.

Work on Waste (WOW) 1. *Auburn, Maine, Incinerator.* 45 minutes, 1985.

WOW 2. *Rome, New York, Incinerator.* 30 minutes, 1986.

WOW 3. *Saugus, Massachusetts, Incinerator.* 34 minutes, 1986.

WOW 4. *Interview with Bernd Franke.* 27 minutes, 1986.

WOW 5. *Windham, Connecticut, Incinerator.* 50 minutes, 1986.

WOW 6. *Recycling in Germany.* 60 minutes, 1986.

WOW 7. *Recycling in the USA: Don't Take No for an Answer.* 60 minutes, 1987.

WOW 8. *Skamania County, Washington Materials Recovery Facility.* 47 minutes, 1987.

WOW 9. *Joe Garbarino: The Only Way to Go.* 38 minutes, 1987.

WOW 10. *Millie Zantow: Recycling Pioneer.* 48 minutes, 1987.

WOW 11. *Zoo Doo and You Can Too.* 59 minutes, 1988.

WOW 12. This video was not distributed.

WOW 13. *How Rodman, New York, Recycles.* 42 minutes, 1988.

WOW 14. *Calvert City: One of Kentucky's Best Kept Secrets.*
60 minutes, 1988.

WOW 15. *Recycling's Missing Link: Fillmore County, Minnesota.*
31 minutes, 1989.

WOW 16. *Waste Management as if the Future Mattered.*
60 minutes, 1989.

WOW 17. *Florida Burning: An Update on Incineration.* 50 minutes, 1990.

WOW 18. *Hazardous Waste Incineration: A Scandal in North Carolina.*
31 minutes, 1990.

WOW 19. *The Differences Between the Theory and Practice of Hazardous Waste Incineration.* 53 minutes, 1990.

WOW 20. *A Regional Medical Waste Incinerator in Hampton, South Carolina.* 52 minutes, 1990.

WOW 21. *Geothermal: A Risky Business in Hawaii's Wao Kele O Puna Rainforest.* 59 minutes, 1990.

WOW 22. *Europeans Mobilizing against Trash Incineration.*
50 minutes, 1990.

WOW 23. *A Hazardous Waste Incinerator in Roebuck, South Carolina.*
25 minutes, 1990.

WOW 24. *Hazardous Waste Incineration in Disguise: Sham Recycling.*
30 minutes, 1991.

WOW 25. *Two Views of Hazardous Waste Incineration in Biebesheim, Germany.* 47 minutes, 1991.

WOW 26. *Community Composting in Zurich, Switzerland.*
50 minutes, 1991.

WOW 27. *Warren County's Incinerator: The Wrong Model for New Jersey.*
39 minutes, 1991.

WOW 28. *Selling Incineration Behind Closed Doors.* 51 minutes, 1991.

WOW 29. *Victims of Incineration in Madison Heights, Michigan.*
59 minutes, 1991.

WOW 30. *Incineration a la Monty Python.* 21 minutes, 1991.

WOW 31. *Wastewise: A Community Reuse and Repair Center.*
29 minutes, 1991.

The following videotapes were produced by Paul Connett for Grass Roots and Global Video (GG Video). All the videos have now been transferred to DVD and copies can be obtained by contacting Paul Connett at 104 Walnut Street, Binghamton, NY 13905, or pconnett@gmail.com. Several are available online at www.americanhealthstudies.org, as indicated below.

GG 1. Oak Ridge: A Case of Broken Faith. 65 minutes, February 1997.

GG 2. May 8, 1997, A Day of Shame for the US EPA—The Hazardous Waste Incinerator in East Liverpool, Ohio. 127 minutes, July 1997.

GG 3. August 14, 1997, A Day of Reckoning for the US EPA—The Hazardous Waste Incinerator in East Liverpool, Ohio. 130 minutes, August 1997.

GG 7. Aloes Weeping: A Case of Environmental Injustice in South Africa. 41 minutes, October 1998.

GG 11. Treading Lightly in Waterloo but "No Standing" in Seneca Meadows, New York. 34 minutes, May 1999.

GG 14. Zanker Road Landfill: A Different Kind of Landfill. 24 minutes, August 1999.

GG 15. American Soil Products: Helping Things to Grow in Berkeley. 29 minutes, August 1999.

GG 16. Zero Waste: Idealistic Dream or Realistic Goal? 58 minutes, September 1999 (available online).

GG 18. The Dangers of Burn Barrels. 27 minutes, November 2000.

Target Zero Canada. 52 minutes, January 2001.

Pieces of Zero, Collection One: Creativity and Leadership. 29 minutes, June 2003 (available online).

On the Road to Zero Waste, Part 1: Nova Scotia. 29 minutes, October 2001 (available online).

On the Road to Zero Waste, Part 2: Burlington, Vermont. 29 minutes, June 2003 (available online).

On the Road to Zero Waste, Part 3: Canberra, Australia. 29 minutes, July 2004 (available online).

On the Road to Zero Waste, Part 4: San Francisco. 29 minutes, 2005 (available online).

Organizations

The California Resource Recovery Association (CRRA) was founded in 1974 and is the oldest and one of the largest nonprofit recycling organizations in the United States. www.crra.com.

The Center for Health, Environment, and Justice (CHEJ) was founded by Lois Gibbs (of Love Canal fame) under a different name, Citizens' Clearing House for Hazardous Waste (CCHW). Their mission is "mentoring a movement, empowering people, preventing harm." For over thirty years this group has helped citizens' groups fight off incinerator and landfill proposals in their communities. http://chej.org/.

Compostable Organics Out of Landfills (COOL) was founded by BioCycle Magazine, GrassRoots Recycling Network, and Eco-Cycle. www.cool2012.com/.

The Container Recycling Institute (CRI) has the mission "to make North America a global model for the collection and quality recycling of packaging materials. We envision a world where no material is wasted, and the environment is protected. We succeed because companies and people collaborate to create a strong, sustainable domestic economy." www.container-recycling.org/.

Cradle2 Coalition is a group of people promoting cradle-to-cradle solutions for zero waste through a national alliance of public interest organizations working for zero waste through green design, source reduction, and producer responsibility for recycling products and packaging. www.cradle2.org/.

Durham Environment Watch (DEW) is a nonprofit and nonpartisan group working to ensure that the York-Durham (Ontario) region becomes a sustainable, healthy community through education and promotion of a zero waste model to conserve valuable resources and promote human and environmental health. www.durhamenvironmentwatch.org/.

Eco-Cycle is recycling operation that serves the city of Boulder, Colorado. Eco-Cycle runs a very important and informative website on the practicalities of its local operation as well as the larger vision of the zero waste movement worldwide. www.ecocycle.org.

The Earth Resource Foundation (ERF) was founded in 1999 as "an environmental educational non-profit organization developed to empower the general public with the resources needed to make environmentally sustainable choices and changes." Its mission statement says that it aims "to preserve, conserve, and restore the Earth to a healthy and sustainable state by redirecting available human, technological, monetary and academic resources." Its website has presentations posted from past zero waste business conferences they supported at www.earthresource.org/zerowaste.html.

Global Alliance for Incinerator Alternatives (GAIA) was founded in December 2010 at a meeting held in South Africa, with the participation of more than eighty people from twenty-three countries. Since this meeting, GAIA has grown to include more than 650 members in over ninety countries. www.no-burn.org.

GrassRoots Recycling Network (GRRN) was founded in 1995 by representatives of the Sierra Club, California Resource Recovery Association, and Institute for Local Self-Reliance. Bill Sheehan was network coordinator and later executive director. www.grrn.org.

Highlights of GRRN zero waste programs, policies, and examples of success stories are at www.grrn.org/zerowaste/index.html.

The GRRN has held three national zero waste conferences. The program for the 2009 conference has some of the presentations available for downloading at www.grrn.org/conference2009/general/agenda.php.

The GRRN website for local government actions provides five fact sheets that detail cutting-edge solutions to urgent waste problems and offer model local government resolutions for communities wanting to take action at www.grrn.org/localgov/index.html.

The GRRN website also highlights businesses around the country that have diverted more than 90 percent of their wastes from landfilling and incineration at www.grrn.org/zerowaste/business/profiles.php.

GRRN's zero waste business principles highlight comprehensive waste diversion goals for businesses to achieve at www.grrn.org/zerowaste/business/.

The Institute for Local Self-Reliance (ILSR) was founded in 1974 in the Adams Morgan neighborhood of Washington, DC. ILSR's mission was to make this community of 25,000 residents self-reliant. ILSR developed programs for waste utilization, urban food production, and decentralized energy systems. www.ilsr.org.

The Product Policy Institute (PPI) is working to transform our throwaway society into a cradle-to-cradle society by developing and promoting extended producer responsibility policies as a prerequisite for zero waste. htttp://www.productpolicy.org.

ReSource (formerly Recycle North) is a large not-for-profit recycling, reusing, repairing, deconstructing, and job training operation located in Burlington, Vermont. http://www.resourcevt.org/.

Samdrup Jongkhar Initiative (SJI) in Bhutan has a delightfully creative "how-to" zero waste pamphlet from Bhutan. It can be accessed at http://www.sji.bt/assets/PDFs/News/zero-waste-manual-small.pdf. For more details of this program please contact Pia Lindström, Zero Waste

Coordinator, Samdrup Jongkhar Initiative (SJI) Dewathang, Samdrop Jongkhar, Bhutan. Pia Lindstorm's email address is pia@sji.bt.

Sound Resource Management Group (SRMG), a consulting company, was set up in Seattle by Jeffrey Morris and others to provide economic analysis on waste and other issues facing local governments. Since then it has been working to shrink pollution footprints, reduce waste, and conserve resources throughout the United States and Canada. The stated goal on its website today is "to help you achieve sustainability and zero waste in the 21st century." In 1992, SRMG played a prominent role in the design of the Nova Scotia waste program. www.zerowaste.com/pages/about.htm.

Urban Ore was begun in 1980 by Daniel Knapp as a salvaging operation at the Berkeley Landfill. Eventually this became a one of the world's largest and most successful reuse operations with the stated mission "to end the age of waste." http://urbanore.com.

Work on Waste USA, the group that eventually became WOW USA, was originally called the National Coalition Opposed to Mass Burn Incineration and for Safe Alternatives. The initial group was formed in 1986. www.americanhealthstudies.org.

The Zero Waste International Alliance (ZWIA) main website is at www.zwia.org. This website includes the only internationally accepted, peer-reviewed definition of *zero waste* to cite in policy discussions (www.zwia.org/standards.html), a list of communities internationally that have adopted zero waste as a goal (www.zwia.org/zwc.html), and zero waste business principles and global principles for zero waste communities (www.zwia.org/standards.html).

Zero waste community plans have been adopted to guide communities in the development of new policies, programs, and facilities to achieve zero waste. Examples of plans that have been adopted include those for Austin, Texas (www.ci.austin.tx.us/sws/zerowaste.htm), and Oakland, California (www.zerowasteoakland.com/AssetFactory.aspx?did=2123).

The number of national zero waste groups continues to grow. Here are some of the more important ones:

> Zero Waste Alliance UK. www.zwallianceuk.org/.
> Zero Waste Canada. www. zerowastecanadanow.ca/.
> Zero Waste Europe. www.zerowasteeurope.eu/.
> Zero Waste Italy. www.zerowasteitaly.blogspot.com/.
> Zero Waste Philippines. www.facebook.com/ZeroWastePH.

INDEX

Note: Page numbers in *italics* refer to figures and photographs. Page numbers followed by *t* refer to tables.

ABOUT THE AUTHOR

Jacci Conderacci Farlow

Dr. Paul Connett, coauthor of *The Case Against Fluoride*, is the director of the Fluoride Action Network (FAN) and the executive director of its parent body, the American Environmental Health Studies Project (AEHSP). He has spoken and given more than 2,000 presentations in forty-nine states and sixty countries on the issue of waste management. He holds a bachelor's degree from the University of Cambridge and a PhD in chemistry from Dartmouth College and is a retired professor of environmental chemistry and toxicology at St. Lawrence University. He lives in Binghamton, New York.

ABOUT THE FOREWORD AUTHOR

Jeremy Irons, actor, is the executive producer of *Trashed*, an award-winning documentary written and directed by Candida Brady, in which Irons travels around the world to record the threats that waste poses to the environment, our health, and the food chain. Irons has won numerous awards for his work as an actor in film, television, and theater, including an Academy Award, as well as Golden Globe, Primetime Emmy, Tony, and Screen Actors Guild awards.

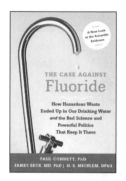

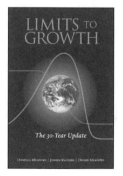